Environmental and Health Impact of Solid Waste Management Activities

ISSUES IN ENVIRONMENTAL SCIENCE AND TECHNOLOGY

TITLES IN THE SERIES:

1 Mining and its Environmental Impact
2 Waste Incineration and the Environment
3 Waste Treatment and Disposal
4 Volatile Organic Compounds in the Atmosphere
5 Agricultural Chemicals and the Environment
6 Chlorinated Organic Micropollutants
7 Contaminated Land and its Reclamation
8 Air Quality Management
9 Risk Assessment and Risk Management
10 Air Pollution and Health
11 Environmental Impact of Power Generation
12 Endocrine Disrupting Chemicals
13 Chemistry in the Marine Environment
14 Causes and Environmental Implications of Increased UV-B Radiation
15 Food Safety and Food Quality
16 Assessment and Reclamation of Contaminated Land
17 Global Environmental Change
18 Environmental Impact of Solid Waste Management Activities

FORTHCOMING:

19 Sustainability and Environmental Impact of Renewable Energy

How to obtain future titles on publication

A subscription is available for this series. This will bring delivery of each new volume immediately upon publication and also provide you with online access to each title via the Internet. For further information visit www.rsc.org/issues or write to:

Sales and Customer Care Department, Royal Society of Chemistry, Thomas Graham House, Science Park, Milton Road, Cambridge CB4 0WF, UK

Registered Charity Number 207890

Telephone: +44 (0) 1223 432360
Fax: +44 (0) 1223 426017
Email: sales@rsc.org

ISSUES IN ENVIRONMENTAL SCIENCE
AND TECHNOLOGY

EDITORS: R. E. HESTER AND R. M. HARRISON

18

Environmental and Health Impact of Solid Waste Management Activities

RS•C
ROYAL SOCIETY OF CHEMISTRY

ISBN 0-85404-285-7
ISSN 1350-7583

A catalogue record for this book is available from the British Library

Published by The Royal Society of Chemistry, Thomas Graham House, Science Park, Milton Road, Cambridge CB4 0WF, UK

Registered Charity Number 207890

For further information see our web site at www.rsc.org

Typeset in Great Britain by Vision Typesetting, Manchester
Printed and bound by Bookcraft Ltd, UK

Preface

Waste management has become a major problem for industrialised societies. It is not that the technologies do not exist; they do, and have done for many years. The main issue is that of public acceptability. The public expect to be able to produce household waste in a largely uncontrolled manner and are accustomed to an efficient local service of removal. Beyond this is where the problems begin. Once the waste is removed from the premises of the producer, most members of the public then regard waste as something that others have produced which should not be treated or disposed of in their locality. Consequently, almost every proposal for a waste disposal facility creates massive public resistance which is heavily exploited by environmental pressure groups with their own agendas.

In an ideal world there would be no such thing as waste, merely useful raw materials for recycling. The reality is of course rather different. Whilst some materials are well suited to recycling, many are not. For those that are suitable for recycling, some kind of segregation at the level of the householder is needed, but this in many countries is not provided. The result is a heavily mixed waste which is most cheaply disposed by tipping in the nearest landfill, which is probably the least environmentally sound option. Recycling such mixed waste involves segregation processes which carry health risks to workers and most probably also to the general public outside of the plant. Even after segregation, there are major waste streams requiring composting, combustion or landfill, all of which carry their own environmental risks.

The topic of solid waste management has therefore become shrouded in emotion, with local government seeking to minimise disposal costs whilst, in the UK at least, doing rather little to meet recycling targets, and environmental pressure groups behaving as if 100% of waste is recyclable and therefore ultimate disposal options such as incineration and landfill are wholly unnecessary and undesirable. This volume of *Issues* is concerned not with the optimisation of solid waste management in terms of recycling and disposal, but rather with the environmental and human health impacts of the various management options.

The first chapter by Kit Strange sets the scene by outlining the various options and broadly addressing their advantages and disadvantages. This is followed by chapters looking at individual waste management options. The first by Dr Toni Gladding of the Open University deals with health risks of materials recycling facilities, highlighting the fact that when such facilities are supplied with mixed waste, they can represent a significant health risk. A substantial quantity of putrescible materials such as food waste have to be disposed of, and can potentially be composted. Composting has been used as a disposal option over many years, but is now increasing in use and is subject to closer scrutiny in relation to environmental emissions. In the following chapter Dr Jillian Swan and Dr Brian Crook of the Health and Safety Laboratory and Dr Jane Gilbert of the Composting Association review the microbial emissions from composting plants.

Historically, landfill sites have led to substantial contamination of the environment. This arises by leaching of contaminants into groundwater, which is mobile, and releases to atmosphere. Modern landfills are altogether better designed and controlled, but concern remains over potential health effects of landfill sites. Unfortunately, most of the research into possible health effects of landfills has not sought to discriminate according to the age and technology of the site. Also, some landfills accept hazardous waste for co-disposal with domestic refuse and these have not been clearly identified in many studies. Drs Andy Redfearn and Dave Roberts of the Applied Environmental Research Centre review studies on the health effects of landfill sites and point to some of the weaknesses. In the following chapter, Professor Paul Williams of University of Leeds reviews environmental emissions from solid waste management activities. These are the starting point for reviewing the impact of chemical emissions on the environment, and in the following chapter, Dr Ari Rabl and Dr Jo Spadaro of the School of Mines, Paris, France, describe how the health impacts of atmospheric emissions from waste incineration may be quantified, and show some of the results of quantification activities. Much of the current understanding of the health impacts of waste disposal is based on the application of epidemiological methodology, which in an ideal world would be supplied with high quality population exposure estimates. Unfortunately, there are many weaknesses to these studies and Professor Helen Dolk of the University of Ulster addresses the methodological issues relating to health effects studies.

As noted above, issues of solid waste management can be highly emotive and all too often are driven by public pressures rather than dispassionate scientific debate. It is hoped that the chapters in this volume will make a strong contribution to scientific knowledge in the area and will be of value to scientists and policy-makers alike.

Roy M. Harrison
Ronald E. Hester

Contents

**Overview of Waste Management Options: Their Efficacy
and Acceptability**   1
Kit Strange

  1 Introduction   1
  2 Key Issues   1
  3 What Is Waste?   2
  4 How Is Waste Managed?   7
  5 Recycling   33
  6 What Is Integrated Waste Management?   34
  7 Trends in Waste Management   39
  8 Public Attitudes   43
  9 Conclusions   50

Health Risks of Materials Recycling Facilities   53
Toni Gladding

  1 Introduction   53
  2 Previous Research Concerning Waste Handsorting   58
  3 A Further Study on MRFs   67
  4 Conclusions   71
  5 Acknowledgements   72

Microbial Emissions from Composting Sites   73
Jillian R. M. Swan, Brian Crook and E. Jane Gilbert

  1 Introduction   73
  2 Bioaerosol Components   78
  3 Potential Ill Health Effects among Compost Workers   84
  4 Compost Site Case Studies   86

Issues in Environmental Science and Technology, No. 18
Environmental and Health Impact of Solid Waste Management Activities
© The Royal Society of Chemistry, 2002

Contents

5   Ill Health Case Studies   96

Health Effects and Landfill Sites   **103**
Andy Redfearn and Dave Roberts

1   Introduction   103
2   Overview of Epidemiological Studies   104
3   Adverse Birth Outcomes   116
4   Theoretical Basis of Purported Effects   129
5   Summary   139

Emissions from Solid Waste Management Activities   **141**
Paul T. Williams

1   Introduction   141
2   Waste Landfill   141
3   Incineration   153
4   Other Waste Treatment Processes   165

Health Impacts of Waste Incineration   **171**
Ari Rabl and Jo V. Spadaro

1   Introduction   171
2   Emissions   174
3   Dispersion and Peak Concentration   179
4   Health Impacts and Costs   182
5   Monetary Valuation   185
6   Calculation of Damage   186
7   Damage Costs per Kilogram of Pollutant   189
8   Damage Costs per Kilogram of Waste   189
9   Conclusions   192
10 Acknowledgements   193

**Methodological Issues Related to Epidemiological Assessment of Health
Risks of Waste Management**   **195**
Helen Dolk

1   Introduction   195
2   The Design of an Epidemiological Study   196
3   Disease/Health Measurement   197
4   Exposure Measurement   199
5   Confounding   204
6   Statistical Considerations   207
7   Criteria for Causation   209
8   Use of Epidemiological Studies in Risk Management   210

Subject Index   **211**

Editors

Ronald E. Hester, BSc, DSc(London), PhD(Cornell), FRSC, CChem

Ronald E. Hester is now Emeritus Professor of Chemistry in the University of York. He was for short periods a research fellow in Cambridge and an assistant professor at Cornell before being appointed to a lectureship in chemistry in York in 1965. He was a full professor in York from 1983 to 2001. His more than 300 publications are mainly in the area of vibrational spectroscopy, latterly focusing on time-resolved studies of photoreaction intermediates and on biomolecular systems in solution. He is active in environmental chemistry and is a founder member and former chairman of the Environment Group of the Royal Society of Chemistry and editor of 'Industry and the Environment in Perspective' (RSC, 1983) and 'Understanding Our Environment' (RSC, 1986). As a member of the Council of the UK Science and Engineering Research Council and several of its sub-committees, panels and boards, he has been heavily involved in national science policy and administration. He was, from 1991 to 93, a member of the UK Department of the Environment Advisory Committee on Hazardous Substances and from 1995 to 2000 was a member of the Publications and Information Board of the Royal Society of Chemistry.

Roy M. Harrison, BSc, PhD, DSc (Birmingham), FRSC, CChem, FRMetS, Hon MFPHM, Hon FFOM

Roy M. Harrison is Queen Elizabeth II Birmingham Centenary Professor of Environmental Health in the University of Birmingham. He was previously Lecturer in Environmental Sciences at the University of Lancaster and Reader and Director of the Institute of Aerosol Science at the University of Essex. His more than 300 publications are mainly in the field of environmental chemistry, although his current work includes studies of human health impacts of atmospheric pollutants as well as research into the chemistry of pollution phenomena. He is a past Chairman of the Environment Group of the Royal Society of Chemistry for whom he has edited 'Pollution: Causes, Effects and Control' (RSC, 1983; Fourth Edition, 2001) and 'Understanding our Environment: An Introduction to Environmental Chemistry and Pollution' (RSC, Third Edition, 1999). He has a close interest in scientific and policy aspects of air pollution, having been Chairman of the Department of Environment Quality of Urban Air Review Group and the DETR Atmospheric Particles Expert Group as well as a member of the DEFRA Expert Panel on Air Quality Standards. He is currently a member of the DEFRA Air Quality Expert Group, the DEFRA Advisory Committee on Hazardous Substances and the Department of Health Committee on the Medical Effects of Air Pollutants.

Contributors

B. Crook, *Health and Safety Laboratory, Broad Lane, Sheffield S3 7HQ, UK*

H. Dolk, *School of Health Sciences, University of Ulster at Jordanstown, Shore Road, Newtownabbey, Co. Antrim BT37 0QB, UK*

E. J. Gilbert, *Composting Association, Avon House, Tithe Barne Road, Wellingborough, Northants. NN8 1DH, UK*

T. Gladding, *Department of Environmental and Mechanical Engineering, The Open University, Walton Hall, Milton Keynes, Bucks. MK7 6AA, UK*

A. Rabl, *Centre d'Energetique, Ecole des Mines, 60 boul. St Michel, F-75272 Paris, France*

A. Redfearn, *Applied Environmental Research Centre Limited, Tey Grove, Elm Lane, Feering, Colchester, Essex CO5 9ES, UK*

R. D. Roberts, *Applied Environmental Research Centre Limited, Tey Grove, Elm Lane, Feering, Colchester, Essex CO5 9ES, UK*

J. V. Spadaro, *Centre d'Energetique, Ecole des Mines, 60 boul. St Michel, F-75272 Paris, France*

K. Strange, *Warmer Bulletin, 1st Floor, The British School, Otley Street, Skipton, North Yorkshire BD23 1EP, UK*

J. R. M. Swan, *Health and Safety Laboratory, Broad Lane, Sheffield S3 7HQ, UK*

P. T. Williams, *Department of Fuel & Energy, The University of Leeds, Woodhouse Lane, Leeds LS2 9JT, UK*

Overview of Waste Management Options: Their Efficacy and Acceptability

KIT STRANGE

1 Introduction

Managing solid wastes in society has been a challenge for as long as people have gathered together in sufficient numbers to impose a stress on local resources. In bygone centuries (and nowadays in poorer countries) waste from homes and industries could be dealt with simply by hauling it to crude dumps where it could be buried, eaten by animals and burned.

Household waste and other waste streams needed to be removed from the human environment to avoid nuisance and public health problems, and the wider environment provided an ample sink for these negative effects of human life. Growth in population and in individual prosperity have since combined to put greater pressure on the environment, at the same time as permitting a growth in people's appreciation of that environment. Consequently waste management policy and practice in industrialised countries developed rapidly in the second half of the 20th century, to ensure that, while public and occupational health risks are minimised, environmental resources are protected.

Since the 1980s one of the driving forces in shaping waste policy, along with many other aspects of life in society, has been the over-arching goal of sustainable development. Within this broader ambition, approaches to waste management have changed, embracing the social, economic and environmental dimensions of sustainability. In practical terms, sustainable waste management has been equated with integrated waste management, the judicious application of a range of options to achieve a broadly optimal system of waste management and resource recovery.

2 Key Issues

The main challenge for a modern, industrial country is to break the historic link between waste creation and wealth creation. Over the years, *per capita* waste

Issues in Environmental Science and Technology, No. 18
Environmental and Health Impact of Solid Waste Management Activities
© The Royal Society of Chemistry, 2002

arisings and wealth (expressed as GDP) have appeared to grow inexorably – with waste production outstripping economic growth. Total reported waste generation within the EU and the European Free Trade Area increased by 10% between 1990 and 1995. Over this period, economic growth was 6.5% in constant prices. The EEA has demonstrated a close correlation between economic activity and municipal waste generation.[1]

Although limited data hinder the development of projections for future waste trends, it is considered that most waste streams will probably increase over the next decade.

The European Union's 5th Environment Action programme *Towards Sustainability*[2] set a target of stabilising municipal waste generation at the 1985 level of 320 kg/capita/annum by 2000. However, there has been little progress and it is clear that this was not met. In fact MSW generation within the EU averaged 400–450 kg per person in 2000, representing a growth of about 30% between 1985 and 2000 (an average annual growth of about 2%). The European Union's 6th Environment Action programme *Environment 2010: our future, our choice*[3] changed from capping individual waste generation rates to setting targets for landfill diversion. This is arguably easier to achieve and is certainly easier to measure.

It remains very difficult to improve the way society uses resources, improving efficiency and reducing the environmental impacts associated with the flow of unwanted materials and energy. This is not because of any particular technical barriers, but is rather a matter of costs and acceptability. While holes in the ground are relatively abundant, more sustainable waste management options can seem unreasonably expensive alternatives, particularly when there is no agreement over who should bear those costs.

3   What Is Waste?

Definitions

The report focuses generally on *municipal solid waste* (MSW). The definition of MSW varies, but typically includes waste arising from private households to that collected by or on behalf of local authorities from any source. MSW therefore includes a proportion of commercial and non-hazardous industrial waste. Depending on the country, the definition can include some or all of:

- household wastes (collected waste, waste collected for recycling and composting, and waste deposited by householders at household waste disposal sites)
- household hazardous wastes
- bulky wastes derived from households
- street sweepings and litter
- parks and garden wastes

---

[1] EEA, *Environmental signals 2001 – European Environment Agency regular indicator report (2001)*, European Environment Agency, Kongens Nytorv 6, DK-1050 Copenhagen K, Denmark.

[2] European Commission (19NN), *European Union's 5th Environment Action Programme Towards Sustainability*.

[3] European Commission (19NN), *European Union's 6th Environment Action Programme Environment 2010: our future, our choice*.

• wastes from institutions, commercial establishments and offices

In Britain, municipal waste is defined as waste collected by, or on behalf of, local authorities and includes all the waste types listed above, although the trade waste component tends to be limited. In most countries, municipal waste is taken to be a broader and more encompassing definition than simply household solid waste.

Depending on definitions, a study by the Dutch Environment Ministry (VROM), in 1997, found that the ratio of household waste to municipal waste varies dramatically from 45% (Norway) to 84% (Germany).[4] It should be recognised that municipal waste is a management concept and since the extent of municipal collection activities in the commercial sector varies widely across the countries surveyed, data and information are unlikely to be directly comparable in a number of cases.

MSW is a small fraction of all solid wastes. The UK generates more than 400 million tonnes per annum (Mtpa) of solid waste, in the following approximate proportions:

• Agriculture	19%
• Mining and quarrying	18%
• Construction and demolition	17%
• Industrial	16%
• Dredged spoils	12%
• Sewage sludge	8%
• MSW	7%
• Commercial	4%

In many ways MSW is the most significant waste stream because it stems from us all, reflecting our lifestyle choices, our consumption and resource recovery decisions.

Waste policies have tended by rely on setting targets which oblige countries or local authorities to recover or recycle specified fractions of waste, or to divert material from disposal to landfill. Inevitably this approach requires decisions on the relative merits of the waste management alternatives. This has become established as the *waste management hierarchy*, a formal order of preference which guides policy-makers.

The Waste Management Hierarchy

Waste management is a complex subject, made up of many component parts. It can be easy to lose sight of the 'big picture'. European waste management is particularly challenging: environmental protection must be achieved without distorting the European internal market. There is no blueprint which can be applied in every situation but the EU has firm principles upon which its approach to waste management is based.

• *prevention principle* – waste production must be minimised and avoided where possible

---

[4] ODEA, *European Commission, DGXI, Statistics on Waste – Phase III, Final Report*, Undertaken by ODEA Consortium (OVAM, DGNRE, ERM and ACR), October 1997.

- *producer responsibility* and *polluter pays principle* – those who produce the waste or contaminate the environment should pay the full costs of their actions
- precautionary principle – we should anticipate potential problems
- proximity principle – waste should be disposed of as closely as possible to where it is produced

These principles are made more concrete in the 1996 EU general strategy on waste which sets out a preferred hierarchy of waste management operations:

1. prevention of waste
2. recycling and reuse
3. optimum final disposal and improved monitoring

The strategy also stresses the need for:

- reduced waste movements and improved waste transport regulation
- new and better waste management tools (*e.g.* regulatory and economic instruments)
- reliable and comparable statistics on waste
- waste management plans
- proper enforcement of legislation

Nowadays it is more often recognised that, although prevention is clearly the ideal option, there are sound reasons why it is not always sensible to specify precisely an order of waste management preferences. A rigid approach has limitations:

- the hierarchy has little scientific or technical basis (there is no scientific reason, for example, why materials recycling should always be preferred to energy recovery)
- the hierarchy is of little use when a combination of options is used (the hierarchy cannot predict, for example, whether biological treatment combined with thermal treatment of the residues would be preferable to materials recycling plus landfilling of residues)
- the hierarchy does not address costs (and so cannot help assess affordability)

However, the hierarchy can be used within an IWM philosophy as a reminder of the waste management options available to the decision-maker.

Composition

The composition of MSW is variable, depending on a range of factors. Household waste reflects population density and economic prosperity, seasonality, housing standards and the presence of waste minimisation initiatives (for example home composting). The prevalence of open fires in the home will affect the levels of combustible materials in the waste stream. Commercial waste will be influenced by the nature of the commerce. The composition of MSW will also depend on the specific definition of MSW being applied. The Resource Recovery Forum[5]

[5] RRF, *Assessment of kerbside collection schemes for dry recyclables*, Resource Recovery Forum, 1st Floor, The British School, Otley Street, Skipton, North Yorkshire BD23 1EP, UK, 2001.

Table 1 Composition on household recyclables and refuse (%, Eastleigh, April 2001)

	Recyclables	Refuse	Total arisings
Newspapers	35.20	3.22	13.01
Magazines	23.98	3.04	9.45
Recyclable paper	5.07	1.95	2.91
Card and paper packaging	7.98	3.33	4.75
Cardboard	4.92	0.58	1.91
Card non-packaging	1.01	0.29	0.51
Liquid cartons	0.47	0.42	0.44
Non-recyclable paper	2.55	6.49	5.28
Refuse sacks and carrier bags	0.59	4.64	3.40
Film: packaging	0.84	4.20	3.17
Film: non-packaging	0.03	0.49	0.35
PET clear bottles	1.94	0.49	0.93
PET coloured bottles	0.38	0.14	0.21
HDPE clear bottles	1.79	0.34	0.78
HDPE coloured bottles	0.91	0.35	0.52
PVC clear bottles	0.14	0.06	0.08
PVC coloured bottles	0.00	0.00	0.00
Food packaging	0.96	3.19	2.51
Non-food packaging	0.70	1.14	1.01
Other dense plastic	0.71	2.68	2.08
Natural and man-made fibres	0.36	2.62	1.93
Disposable nappies	0.13	5.50	3.86
Shoes	0.00	1.00	0.69
Wood	0.16	1.23	0.90
Other	0.82	2.82	2.21
Unclassified non combustibles	0.05	3.45	2.41
Clear bottles and jars	0.12	2.39	1.69
Green bottles and jars	0.09	0.70	0.51
Brown bottles and jars	0.09	0.61	0.45
Other glass	0.38	3.36	2.45
Food cans	4.02	1.94	2.58
Beverage cans	0.42	0.14	0.23
Batteries	0.01	0.00	0.00
Aerosols	0.15	0.27	0.23
Other ferrous	1.09	1.99	1.71
Aluminium foil	0.03	0.44	0.31
Aluminium beverage cans	0.72	0.29	0.42
Aluminium food cans	0.00	0.05	0.03
Other aluminium	0.29	0.94	0.74
Garden waste	0.00	0.31	0.22
Kitchen compostables	0.07	3.00	2.10
Kitchen non-compostable	0.48	24.35	17.04
Particles < 10 mm	0.35	5.56	3.96
Total	100.00	100.00	100.00
Arisings, kg/household per week	4.5	10.0	14.5

Source: Resource Recovery Forum (2001).[5]

Table 2 Comparison of waste compositions in Europe

Country	Data year	Arisings (tpa × 10³)	Paper & card	Glass	Metals	Textiles	Plastics	Organics	Other
Belgium (Brussels)	1997	337	18	10	3	3	7	31	28
Belgium (Wallonia)	1995	1524	20	9	3	2	8	33	25
Denmark	1985	1900	22	2	3	5	4	55	9
Greece	1996	3606	18	3	3	4	10	51	11
Spain	1996	15 305	21	7	4	5	11	44	8
France	1995	26 000	25	13	4	3	14	29	15
Ireland	1995	1027	23	6	3	3	10	34	21
Luxembourg	1993	99	19	7	3	2	8	43	17
Netherlands	1996	7537	27	6	2	2	5	39	18
Austria	1996	2775	24	9	7	3	15	29	1
Finland	1994	2100	33	2	5	2	3	33	35
Bulgaria	1997	3628	11	6	3	4	7	38	31
Czech Republic	1996	3200	8	4	2	2	4	18	61
Hungary	1996	5000	19	3	4	3	5	32	33
Latvia	1995		14	8	4	3	7	48	16
Lithuania	1997		1	2	19	1	0	40	0
Romania	1997	4357	17	6	6	6	7	56	2
Slovakia	1996	1700	13	6	8	3	9	26	35
Slovenia	1995	1024	15	5	7	0	10	32	31
Norway	1997	1354	31	3	4	5	8	27	22
Switzerland	1994	2660	29	3	3	2	15	38	10

published a report which showed the results of extensive analyses of one source of household waste, showing the detailed breakdown of residual waste and materials recovered for recycling (Table 1).

Eurostat[6] reports that data availability on waste composition is generally very poor and difficult to compare, not least because the characterisation techniques may vary from country to country. Because of this, the method used to assess the weight of the various fractions may influence the results due to the fact that objects containing various materials may or may not be assigned to a single category. The compositions reported (see Table 2) may not always refer to the total amount of municipal waste generated which should – but does not always – include all the waste fractions separately collected for recycling and recovery operations.

Waste Arisings

The rate at which individuals create waste is at least as variable as the composition of the waste generated. The average European now creates more

---

[6] Eurostat, *Regional environmental statistics – initial data collection results*, ISBN 92-828-6259-3, 2001.

Figure 1 Municipal waste generated per capita by country in western Europe

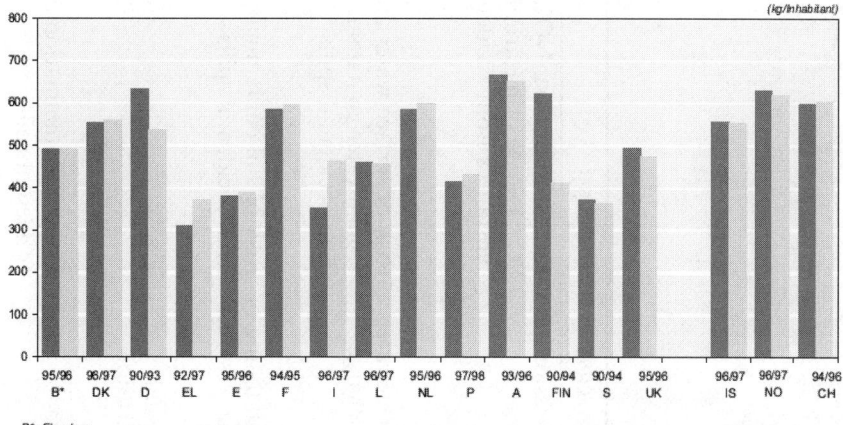

(kg/Inhabitant)

B*: *Flanders*

than half a tonne of municipal solid waste each year, and arisings are generally stable, although growth is still taking place in some countries.

The European agency Eurostat publishes waste and other environmental statistics, which indicate[7] that the trend in municipal solid waste arisings continues to grow, accounting for approximately 190 million tonnes per annum (Mtpa) in the mid-1990s, averaging 575 kg per person per annum (see Figure 1).

4   How Is Waste Managed?

Essentially, the final management option for residual waste is disposal to landfill. Everything else might be viewed as some form of pre-treatment or waste avoidance. However, it is clear that all of the alternative management options have a role to play, to help minimise the amount of material which can be said to have been wasted. Indeed, some optimistic policy-makers do not regard even landfill disposal as an ultimate waste of resources, preferring to regard this as a long-term storage element of a wider *Zero waste* policy.

The range of options available once waste is created is limited to recovering materials and/or energy resources before the final, useless residues are landfilled. The techniques applied depend on the materials in the waste, the waste management systems available locally or regional and the market opportunities. The selection depends particularly on the established waste management policy. Treatment methods are used to reduce the amount of residual waste for disposal, and to achieve one or more of the following goals:

• reduce the potential environmental impacts of the waste
• separate and recovery materials or energy
• reduce transport costs
• reduce the
• volume of landfill needed
• minimise overall costs

[7] Eurostat *Waste generated in Europe – data 1985–1997*, ISBN 92 828 7941 0, 2000.

7

Table 3 Environmental impacts of waste management methods

	Landfill	Composting	Incineration	Recycling	Transport
Air	Emissions of methane (CH_4) and carbon monoxide (CO) odours	Emissions of methane (CH_4) and carbon monoxide (CO) odours	Emissions of SO_2, NO_x, HCl, HF, NMVOC, CO, CO_2, N_2O, dioxins, furans, heavy metals (Zn, Pb, Cu, As)	Emissions of dust	Emissions of dust, NO_x, SO_2, release of hazardous substances from accidental spills
Water	Leaching of salts, heavy metals, biodegradable and persistent organics to groundwater	N/a	Deposition of hazardous substances on surface water	Wastewater discharge	Risk of surface water and groundwater contamination from accidental spills
Soil	Accumulation of hazardous substances in soil	N/a	Landfilling of ashes and scrap	Landfilling of final residues	Risk of soil contamination from accidental spills
Landscape	Soil occupancy; restriction on other land uses	Soil occupancy; restriction on other land uses	Visual intrusion; restriction on other land uses	Visual intrusion	Traffic
Ecosystems	Contamination and accumulation of toxic substances in the food chain	Contamination and accumulation of toxic substances in the food chain	Contamination and accumulation of toxic substances in the food chain	N/a	Risk of contamination from accidental spills
Urban areas	Exposure to hazardous substances	N/a	Exposure to hazardous substances	N/a	Risk of exposure to hazardous substances from accidental spills; traffic

Figure 2 Waste management in Europe (Source Eurostat/OECD)

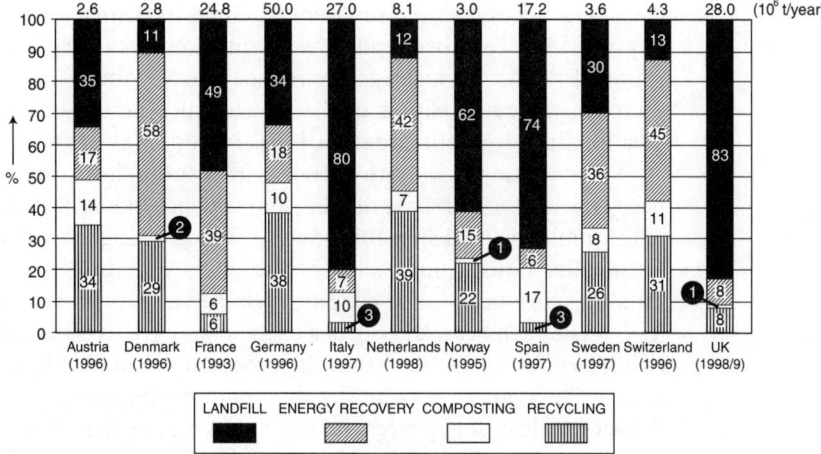

Waste management options after generation and before final disposal comprise:

• waste minimisation
• collection and sorting
• re-use
• recycling
• composting
• anaerobic digestion
• energy recovery (incineration or other more advanced thermal treatment techniques)
• incineration (without energy recovery)

Arguments over the relative merits of management options between prevention and disposal have been loud and vigorous, although during the late 1990s there were significant signs of rapprochement between the different camps. Although local circumstances may alter the priorities, it is generally true that the main purpose of a waste management policy is to reduce the scope for waste to harm the environment or public health. The European Commission has described[8] the range of environmental impacts from managing MSW, as shown in Table 3.

It is clear from many sources that different countries elect to manage waste in different ways, or rather display different sets of preferences as alternatives to landfill. A study undertaken for the Resource Recovery Forum[9] yielded the results shown in Figure 2, which shows that landfill dependency ranges by an order of magnitude.

Waste Minimisation

Waste minimisation, prevention or avoidance is the most important management technique to be applied to solid wastes, because waste which is avoided needs no management and has no environmental impact.

[8] European Commission, *Waste management options and climate change*, ISBN 92 894 1733 1, 2001.
[9] RRF, *Recycling achievement in Europe*, Resource Recovery Forum, 1st Floor, The British School, Otley Street, Skipton, North Yorkshire BD23 1EP, UK, 2001.

General recommendations, while helpful in theory, often contribute little in any individual case, over and above giving pointers as to possible waste reduction routes. In order to successfully reduce waste volumes, it is first necessary to establish the composition of that waste, and the reasons which prompted its creation. In a domestic situation, those reasons may include or be dictated by life style, for example if both parents in a household are working full time, necessitating the purchase of more convenience foods, or if there is a young baby in the family using disposable nappies. The growing tendency in some countries for smaller households increases the likely quantities of wastes created. For example one study[10] showed that single-person homes can create around 120 kg food waste each year. By contrast each person in a two-person household creates only 85 kg food waste pa. In a three-person, home this figure falls to 50 kg per person per year. In a commercial situation, some waste may be the result of delivery policies set by a central supply system, or stem from choices made years before on types of machinery, requiring considerable investment in new equipment to change.

Definition.   While some consider the diversion of waste materials into recycling to be waste minimisation, as this reduces the amount of waste going for final disposal, the original intent of the term was to reduce to a minimum the amount of waste being generated. The OECD[11] has developed a broad definition of waste minimisation, encompassing the following three elements, in this order of priority:

- preventing and/or reducing the generation of waste at source
- improving the quality of the waste generated (*e.g.* reducing the hazard)
- encouraging re-use, recycling and recovery

Home composting schemes, where householders themselves turn kitchen and garden wastes into compost, can be considered waste minimisation, but all other procedures which require handling of wastes such as collection for recycling, while reducing pressure on disposal facilities, are not actually reducing the amount of waste generated. That reduction could be achieved through changed manufacturing procedures, or even a change in the type or combination of materials used, enabling production scrap to immediately be returned to the beginning of the manufacturing line, so that it never reaches the stage of being considered a waste.

Resource optimisation.   It is easy to think narrowly of waste as a solid product left over at the end of a process or action, but it would be wrong to concentrate on reducing the amount of solid waste produced, to the exclusion of considerations about, among other things, wastage of energy or water. It is as wasteful to use several gallons of water unnecessarily, or to drive cars when one could walk or cycle, or to consume energy thoughtlessly, as it is to discard newspapers, cans or empty wrappings.

[10] Incpen, *Towards greener households: products, packaging & energy*, Incpen, Tenterden House, 3 Tenterden Street, London W1R 9AH, UK, ISBN 1 901 576 50 7, 2001.
[11] OECD, *Pollution prevention & control: considerations for evaluating waste minimisation in OECD member countries* (Report ENV/EPOC/PPC(978)17/REV2), OECD, Paris, 1998.

For example, it can be wasteful (and environmentally counter-productive) to drive several miles to deposit a few newspapers, empty cans, glass or plastic bottles into collection banks. The resources consumed in doing so, together with the further resources needed to take the materials for reprocessing, could exceed the resources saved by them not being thrown away.

There can be a conflict between minimising the resources used to make a product and the possibility of re-using, or recycling, the product. For example, in order to make a refillable glass bottle strong enough to withstand several trips between consumer and bottling plant, the bottle must be made stronger, which uses more glass. The heavier the container, the more transport needed to carry it, and its contents, to their destination. Transport has a large environmental impact. Therefore there needs to be careful evaluation of the number of return trips each bottle makes set against the increased resource use for making it refillable. If the refillable bottle is thrown away and not returned, or only refilled once or twice, the resources wasted are greater than if the bottle had been designed for a single use. Recycling is aided by the use of fewer different materials in a single product: for example, the recent switch by car manufacturers to use only three or four types of plastic, instead of more than twenty formerly, has simplified the recovery of plastics from scrap cars. Sometimes a combination of paper and plastic, or plastic coating on glass, enables a product to fulfil its role with the minimum resources, but, because of the mixed materials, it is harder to recycle than, say, an all paper item.

It is wasteful to allow food, which has consumed resources and energy in its production, to be damaged or spoiled. Extreme measures designed to reduce packaging may have the effect of reducing the use of paper, metals, glass and plastics at the expense of the food they would have protected, despite the value of the wasted food being many times greater than the value of the now-avoided packaging. Surveys show that householders in industrialised countries waste significant amounts of food, simply through buying too much. It is wasteful to demand new clothes and new furnishings simply to follow fashion trends when the current articles still have useful life left in them.

Waste minimisation at home. Waste minimisation is hard to achieve for individuals and households, but there are some contributions which can be made. For example, care should be taken when purchasing goods that appropriate amounts and sizes are chosen. Buying large tins of paint to do a small decorating job, or buying larger amounts of food than can be consumed while fresh, are two examples of unnecessary waste creation.

Since waste is not simply the refuse put out for collection each week, considerable waste reduction in terms of resource use can be made by using electricity sparingly, by cutting down the number of car journeys made, and so on. Individuals can reduce the amount of waste they create by buying less, by buying longer-life products, and by re-using items: empty tins and jars make good storage, yoghurt pots are ideal for seedlings, magazines once read can be passed to neighbours and friends. Mending broken or worn items of clothing or equipment has a further important contribution to make.

Waste minimisation in industry.   The trend to conduct environmental audits of companies, products and processes is likely to provide pointers to minimisation policies, and will help determine where environmental burdens are imposed. In a trial in Kent, UK, eleven companies were helped to identify a range of waste reduction initiatives. In the initial phase of the study the companies saved money (more than £2 million pa), identified the means to reduce waste arisings sent to landfill (116 000 tonnes pa), found opportunities to reduce water demand (940 000 m^3 pa) and ear-marked energy savings equivalent to carbon dioxide emissions reduction of 1700 tonnes pa).

Another project involved 24 companies in South Yorkshire, UK. These companies identified savings of:

- > £1 million pa; 5100 m^3 of solid waste to landfill
- 275 000 m^3 of water
- 4500 tonnes of carbon dioxide.

In Austria, the government of Styria[12] set out a programme during the 1990s to demonstrate the scope for waste prevention through cleaner production techniques. More than 200 options were identified, with potential savings of ATS20 million pa. The distribution of these opportunities was as follows:

- good housekeeping (20%)
- organisation of data (24%)
- process changes (33%)
- internal recycling (6%)
- external recycling (2%)
- change of raw materials (9%)
- stabilisation (6%)

Waste exchanges, where the waste product of one process becomes the raw material for a second process, are another way of reducing waste disposal volumes for those wastes which cannot be eliminated.

Long life products.   Product durability is a way of reducing waste and, in the majority of examples, extending, say, a vacuum cleaner's useful life to fifteen years instead of twelve, can make a major contribution to resource optimisation.

There are, however, some instances where extending the life of an item could actually be a negative approach, in total environmental terms. If your old washing machine uses double the amount of water, detergent and energy as the newer model, and you use it very frequently, extending its useful life may, overall, place a heavier burden on the environment than sending the old one to be scrapped, its metal recycled, and purchasing a new model. Motor cars are another example: generally, old vehicles consume more fuel and produce more emissions than their modern counterparts.

Product design.   Waste minimisation strategies should start at the beginning of a product's life cycle – with product design stage. In addition, reducing the number

[12] Warmer Bulletin, *Waste minimisation*, Warmer Bulletin Information Sheet, September, 2000.

of different components used in a product, or making it easier to take apart, can make the task of recycling it at the end of its useful life simpler, as well as making its repair possible. There may need to be a choice made as to whether reducing the amount of raw materials to make the goods is of paramount importance. It could be that reducing the volume or toxicity of waste created when the goods have reached the end of their life, or even making the products environmentally less harmful while in use, have higher priority.

Hiring, sharing and borrowing.   There are considerable opportunities for waste reduction in hiring or borrowing equipment which is not often required. Examples of this include garden tools like shredders and hedge cutters, and decorating equipment. Special occasion clothing can also be hired. Shared or communal use of equipment is another waste saving measure. For example a small community or neighbourhood group could jointly purchase a commercial sized washing machine and run their own laundry service, saving on energy and water as well as on the number of pieces of equipment needed. This could be particularly relevant in a community with a large number of babies and young children using nappies.

Waste minimisation instruments.   A number of policy instruments can be regarded as addressing waste minimisation. These include *plans* and *programmes*, *mandatory* instruments (technical standards and product bans), *economic* instruments (taxes and duties, financial aid and economic incentives) and *suasive* instruments (information provision and public relations, environmental management systems, environmental reporting and eco-labelling).

A study by the Institute for Applied Ecology in Darmstadt, Germany[13] concluded that product-oriented environmental policies will be the most effective means of delivering waste minimisation in the longer term. National environmental and waste policies are also changing to reflect these new priorities. For example, the government of Victoria, Australia introduced a ten year strategy *Zeroing in on wastes* in 1998 with a specific objective to achieve '*the widespread avoidance of waste by facilitating the adoption of cleaner production policies and practices*'. In 1999 this was followed by a target to reduce waste to landfill by 10% by 2010.

By being aware of the impacts of purchasing decisions, both industry and individuals can make a difference to total resource consumption.

Waste Collection and Sorting

The way that wastes are collected and sorted influences which waste management options can most effectively be used. The collection method significantly shapes the recovery of materials, compost or energy; this in turn determines whether markets can be found. Collection is also the point of contact between generators (*e.g.* households and commercial establishments) and the waste management system. Collection is rarely independent of subsequent sorting, since the type of

[13] Oko-Institut eV, *Waste prevention and minimisation: final report*. Commissioned by European Commission DG XI from Oko-Institut eV, Darmstadt, Germany, 1999.

collection affects the sorting needs, and some collection methods themselves involve a level of sorting.

Householder sorting. From the householder's viewpoint, commingled collection of all waste is probably the most convenient method, although this limits subsequent treatment options. Most treatment methods need some form of separation of the waste into different fractions in the home, prior to collection. At its simplest this might involve removing recyclable materials.

Household collection systems are often divided into '*bring*' and '*kerbside*' collection schemes. Bring systems are those where householders are required to take recyclable materials to communal collection points. Kerbside collection schemes require the householder to place recyclables in a container which they set, on a specified day, outside their property for collection. The extreme bring system is the central collection or Civic Amenity site to which householders transport materials such as bulky items and garden waste. Such sites often also have collection containers for recyclables. Other bring systems comprise materials banks at low density, often situated at supermarkets.

Kerbside collection is more narrowly defined, but collection can also be of separated fractions or of commingled waste.

Collecting household waste. Household waste has traditionally been collected mixed, but where household sorting has occurred the different waste streams are collected separately, whether by the same or different collection vehicles. The categories collected separately vary by geography: in Germany, the *Duales System Deutschland* (DSD) collects packaging material as a separate stream, whereas in Japan householders separate out combustible material for separate collection. In Europe and North America, separate collections are most commonly used for dry recyclables (paper, metal, glass, plastic), *biowaste* (kitchen and garden waste, with or without paper) and, in some countries, *household hazardous waste* (batteries, paint, *etc.*). A collection for remaining residual waste (known as restwaste) is also needed. Garden waste and bulky waste may be handled as separate streams, or alternatively included within the biowaste and restwaste streams, respectively.

Collecting dry recyclables. The greatest range of collection methods, from central or low density materials banks, to kerbside collection of recyclable materials in specially designed trucks. Single (mono) material banks ('drop-off') that collect a single material per container represent one of the best known forms of materials collection, mainly due to the success of 'bottle banks' for glass.

A range of kerbside collection methods have been used to collect recyclables, varying in the degree of sorting involved, and including boxes, bags and wheeled bins. In its simplest form, recyclables are separated by the household and stored together in a bag, box or wheeled bin ready for collection. As with the collection of mixed recyclables from street containers, collection can use existing collection vehicles, in some cases even with compaction. Commingled collection of recyclables, whether from communal kerbside containers or household bags or bins requires extensive subsequent sorting at a Materials Recovery Facility

(MRF). McDougall *et al.*[14] quote Schweiger (1992) which gives a comparison in recovery rates for kerbside and bring systems in Germany (Table 4) which shows that the former can achieve the highest levels of recovery, although great success can accompany the use of high density bring facilities.

Central sorting. Materials recovery facilities (MRFs) are facilities that process solid wastes for the purpose of recovering commodity-grade materials for sale, or recovering a mixed material fraction for subsequent processing or conversion (for combustion or mechanical-biological treatment). Depending upon the intent and design of the MRF, the wastes may be delivered to the facility as mixed solid wastes, as source-separated wastes (or alternatively, source-separated materials), or in both forms. However, facilities tend to process either mixed wastes or source-separated materials, not both.

The distinction between mixed solid wastes and source-separated materials is not a clear one. Mixed solid waste is heterogeneous, while source-separated materials are composed of specified components targeted for recycling. As more components are added to source-separated collection programmes in an effort to recycle more materials, the source-separated mixtures begin to approach the characteristics of mixed solid waste.

Both general categories of MRFs (*i.e.* those that process mixed wastes or those that process source-separated wastes) can utilise mechanical, manual, or both methods of processing materials. To achieve high levels of diversion of wastes from landfill, both source-separated and mixed waste MRFs may be required for a given community or region. For this discussion, high levels of diversion are those beyond the 40–50% level. The structuring of systems that would use both types of generic MRFs is an exercise in integration of the facilities in terms of compatibility and optimum efficiency and economics.

Mixed waste MRFs, in industry jargon, are also called 'dirty' MRFs, whereas source-separated MRFs are also called 'clean' MRFs. The term 'dirty' refers to the high level of contamination and moisture usually found in mixed solid wastes, while 'clean' describes the relatively uncontaminated materials that are provided by source-separated collection programmes.

Mixed waste MRFs. Mixed waste MRFs are designed to process residential, commercial, institutional, or industrial mixed solid wastes, or any combination of these. Levels of contamination in the delivered wastes are expected to be high, and non-processible items will require special handling. Non-processible items include stringy items such as cable, textiles, and magnetic tape and over-size objects such as exhausts from vehicles, water heaters, and rolled carpet. The gross recovery rate for commodity-grade secondary materials can be expected to be low (*e.g.* 10–30% of the quantities processed) due to the high levels of contamination and to concomitant low yield of products of marketable quality.

An exception can be the processing of mixed solid wastes from a specific type of generator where higher diversion is possible due to large concentrations of non-contaminated material (*e.g.* paper, in the case of commercial businesses).

[14] F. McDougall *et al.*, *Integrated solid waste management: a life cycle inventory*, 2nd Edition, ISBN 0 632 05889 7, 2001.

Table 4 Comparison of German kerbside and bring systems

	Paper		Glass		Metals	Plastics	Mixed dry recyclables	Biowaste
	kg/person/y	Recovery rate (%)	kg/person/y	Recovery rate (%)	kg/person/y	kg/person/y	kg/person/y	kg/person/y
Bring systems								
2000 persons/bank	5–15	8–25	5–15	13–38	0.5–2.5	1–2	15–50	5–30
1000 persons/bank	10–25	17–42	10–20	26–51	0.5–2.5			5–30
500 persons/bank	15–50	25–50	15–25	38–64	0.5–2.5			5–30
Kerbside collection								
paper collected (in bundles):								
Every week	20–35	33–58						
Every 2 weeks	15–25	25–42						
Every 4 weeks	10–20	17–33						
Paper (in containers)	35–55	58–92	15–30	38–77				
Glass (in containers)								
Multi-material (glass, metals, plastics)								
Bag collection	30–50	50–83	12–30	31–77	5–10	5–10	30–60	
Bio-bin	5–25	8–42	5–20	13–51	1–2	5	50–140	

Rates of recovery for mixed waste MRFs processing residential mixed wastes can be expected to be, and usually are, lower than those of facilities processing commercial and industrial mixed wastes. Typically, mixed waste MRFs have processing rates in the range of 200–1500 tonnes per day.

Source-separated MRFs (SSMRFs). Source-separated MRFs are designed to process individual components (*e.g.* aluminium cans), mixtures of individual components (*e.g.* commingled tin, glass, and aluminium containers), or both. Consequently, source-separated MRFs (SSMRFs) can be further divided according to the degree of mixing or commingling of the components, and also the number of processing lines dedicated to processing the different source-separated mixtures. For all SSMRFs, the quality of the materials delivered to the facility is expected to be high and contamination low. Process residues from SSMRFs typically are in the range of 3–10% of the weight processed, although values as high as 15% have been experienced.

General considerations. The primary objective of a MRF is the processing of the raw feedstock into a marketable or usable end-product. The processing must be carried out such that the end-products meet a given set of specifications. Also, especially in times of tight markets, recovered materials of high quality are desirable to assure continuity of end-product sales.

A variety of equipment is used in a materials processing facility. Conveyors (predominantly belt conveyors) are essential pieces of equipment in a MRF. Shredders, crushers, magnetic separators, screens, and densifiers (*e.g.* balers) are also frequently used. Rolling equipment used at MRFs includes front-end loaders, forklifts, and trailers. The design of mixed waste MRFs must incorporate processes to open bags to liberate their contents for segregation and potential recovery. Source-separated MRFs that accept bagged materials also require similar processes.

Regardless of whether a mixed waste or source-separated MRF is employed for recovering recyclable materials, successful design and equipment selection must accommodate the particle size distribution and other pertinent characteristics of the materials to be processed, achieve the necessary level of purity and yield of end-product, and prepare the end-products to the other specifications of the marketplace, including, for example, bulk density.

Composting

The word 'compost' conjures up images of an untidy heap of assorted kitchen and vegetable remains at the far end of the garden. A more scientific definition of compost would be the product of natural degradation of botanical and putrescible waste by the action of bacteria, fungi, insects and animals in the presence of an adequate air supply. The biological decomposition processes break down complex organic substances into carbon dioxide, water and a residue: compost. The final product is relatively stable and can be used as a soil improver, a mulch or a component in a growing medium.

Composting occurs when there is a plentiful supply of oxygen, moisture and warmth. Compost must be regularly aerated, by turning the heap or by injecting air into the composting material, for the process to be successful. Under properly controlled conditions the temperature of composting refuse can rise to levels which are sufficient to kill pests, weed seeds and pathogenic bacteria.

Anaerobic fermentation occurs when the oxygen supply is restricted. Complex hydrocarbons are broken down into reduced intermediate by-products; both methane – which has a potential value as a fuel – and carbon dioxide are released. Anaerobic decomposition – called fermentation or digestion – occurs all the time in landfill sites containing household refuse.

Three main classes of micro-organisms are involved in the decomposition of refuse: bacteria, fungi and actinomycetes. Bacteria and fungi predominate as the organic material begins to decompose. If sufficient air is available, the rate of metabolic activity of these micro-organisms is so great that temperatures can rise to 70 °C or more. At this stage, only heat-tolerant (thermophilic) bacteria and actinomycetes can continue to decompose the waste. Gradually, as the substrate is consumed, the rate of decomposition slows, the compost cools, and the fungi and non-thermophilic bacteria become active again. All micro-organisms need water to function; if the moisture content of composting waste falls below 40%, microbial activity slows. If the moisture content is too high, however, air spaces within the composting material fill with water, creating anaerobic conditions which can cause unpleasant odours.

Mature compost is a valuable substance. It can act as:

- *a soil conditioner* – improving soil structure, especially for heavy clay soils. Compost also retains moisture and so helps to improve light sandy soils. It reduces soil erosion, and helps to bind nutrients, preventing them from being washed out of the soil
- *a soil fertiliser* – encouraging a vigorous root system
- *a mulch* – if applied around plants it will smother small weeds and prevent the surface soil from drying out
- a peat substitute – for use in potting mixtures

Household refuse contains a high proportion of organic materials which are suitable for separation and composting. The compostable fraction of domestic waste includes food scraps, animal wastes, soft plant materials, paper and card. However, not all materials of biological origin decompose fully during composting. Less readily decomposed materials include wood, bone and industrially 'altered' organic materials such as leather. Surveys show that MSW often contains as much as 50% organic material.

A typical aerobic composting system for mixed waste has several stages:

Pre-processing. To modify the incoming material, making it more suitable for composting. Materials recovery takes place at this stage, and the waste is then pulverised and/or screened to separate the organic fraction.

Decomposition. The methods used for the large-scale production of compost fall into three main categories. Windrow composting uses mechanical turning to aerate periodically the composting waste, which is placed in elongated heaps around two metres high. Blowers may be used to force air through the windrows for more efficient aeration and heat removal.

In forced aeration (or static pile) systems, the waste is carefully piled over a ventilated floor area or a perforated pipe, through which air can be passed. The ventilation may be vacuum-induced, with air drawn inwards through the waste. This allows any odours produced to be contained and, if necessary, treated. Alternatively air may be blown outwards through the composting waste; this method is often preferred as it allows the heat from the most rapidly decomposing waste at the centre or base of the pile to be transmitted to the outer, cooler regions. Forced aeration systems generally have higher capital costs, and lower operating costs, than windrow systems. Alternatively, a variety of enclosed reactor systems may be used, which are designed to allow close control over temperature, moisture and aeration and mixing rates, but which also require a high investment of capital. These technologically more advanced systems are typically applied to more complex mixtures of wastes.

Maturation. Even when the compost has been stabilised using one of the above methods, it may not be ready for use. During decomposition, complex organic molecules are converted by stages into simpler compounds. Some of the intermediate breakdown products in this process can be toxic to crops. The compost may need to pass through a final maturation or curing stage while these plant toxins are further decomposed before it can be used. This may take weeks or months.

The amount of compost which a mixed municipal waste composting plant can produce will, of course, vary according to the quantities of waste received, but will also depend on the composition of the waste and the quality required of the final product. Most organic waste has a high moisture content, and there is a substantial drying effect during composting; there is also a considerable loss in dry weight, due to the transformation of organic carbon to gaseous carbon dioxide. The final stages of processing the composted material also have an effect on the yield of a composting plant. For high quality compost, it is usually necessary to incorporate a final sorting stage to remove uncomposted particles and inert contaminants such as wood, glass or plastic. Such final screening stages may not always be necessary, for example in cases where the compost is used for landscaping, land reclamation or landfill cover.

Compost from mixed domestic wastes is not easily marketed. Cases of failure to find markets have been well documented in Germany and other European countries. However, compost produced by plants in the Middle East has been in such demand that in some cases would-be customers have had to join a two year waiting list. Composts produced in Japan (from sewage sludge, agricultural and municipal wastes) and the former USSR are used almost exclusively by farmers. Increasingly, countries are favouring the separate collection of putrescible household wastes. In the Netherlands, separate collection systems have been mandatory since 1994. Austria, Germany and Switzerland all have similar requirements now in place.

Compost must compete with other soil conditioners, and municipal authorities have been reluctant to invest in plants without an assured market for the product. Markets have been hampered by poor quality, badly stabilised compost, with visible contaminants such as pieces of glass and plastic. However, even when no ready market is available, composting can provide an effective means of reducing the polluting potential of household refuse. Composting stabilises degradable materials and reduces their volume, conserving landfill space and decreasing the risk of pollution from landfill gas and leachate. Another objection to municipal compost is that it may contain heavy metals (and other toxic substances) because of its origin as mixed refuse. It is difficult to separate out the main sources of heavy metals before the refuse is composted, and research is currently under way to improve the process. Most of the heavy metals present are found in the smallest particles of the household waste, so effective screening can remove a high proportion. However, complete removal of these contaminants is rarely possible. Some of the most difficult metal-rich contaminants to remove are household dust (which can have a high lead content), staples, wine bottle caps and batteries. To produce high quality composts, it is far better to ensure that contaminants do not mix with the compostable waste, by source separation for central composting, or by composting at home.

If the individual householder composts organic refuse at home, not only is it removed from the waste stream, but the householder has complete control over the type of refuse put on the compost heap, and can therefore include only biodegradable material. Where it is not possible for householders to compost their own refuse, the local authority can operate a service in which the organic fraction is collected separately from other rubbish. In this case, the resulting compost will be of much more consistent quality, and many of the problems (such as contamination with heavy metals) are minimised.

If organic waste is not separately collected, the production of compost from MSW is perhaps best carried out as part of an integrated waste management system. Such a plant has been operating in Duisberg, Germany, for 30 years. Mixed household waste from an area serving 95 000 residents is first passed beneath a magnet to remove ferrous metals, then along conveyor belts where glass bottles, non-ferrous metals and hard plastic items are picked out by hand. The rest of the refuse is then fed into a rotating reactor vessel, in which temperatures and moisture are carefully controlled. The waste remains in this reactor for 36 hours, and is then screened. The large particles (paper, card, textiles and plastic) are incinerated, and the smaller fraction is separated into two different grades of compost which are matured for a further three weeks, aerated with the air which is expelled from the enclosed reactors. The compost is used mainly in vineyards, with most of the rest used for landscaping and land reclamation. In Germany, composting activity has increased considerably in recent years. In 1985, twelve facilities produced around 0.25 million tonnes pa. By 1993, there were around 100 plants in operation, yielding 1.1 million tonnes pa. Almost 100 other projects were either under construction or had received planning approval.

In the 1980s, French composting facilities would traditionally process mixed household solid waste. As a result, the product was of low quality, tainted by heavy metals and containing sharp glass and plastic film. As farmers demanded higher quality material for land application, so end-markets for mixed waste compost declined. Early in the 1990s a few municipalities (Bapaume, Niort, Lille and Le Creusot) introduced selective kerbside collection for compostable waste.

Of these, Bapaume has now developed a cost-effective system that works well in a rural or semi-urban setting. The Bapaume experiment, which collected green wastes, kitchen wastes, disposable nappies and cartons) demonstrated a range of benefits:

- bio-bins collected 40% of the waste stream
- more than 95% of the bio-bin contents were compostable
- less than 1% of bio-bins were refused by collectors

Trials were scaled up into a commercial composting plant using the OTVD-SILODA process. The unit, operational in 1997, will process up to 14 000 tonnes pa as more municipalities join the scheme. A survey in France in 1993 suggested that there were 73 MSW composting units (no source separation), 30 green waste composting plants, 16 farmyard manure composting facilities, ten mixed organics composting plants and five experimental source-separated plants.

In America, the total number of garden waste composting facilities reached more than 3300 in 1995. Some two-thirds of these involve the composting of leaves collected in the Autumn. Around half of US states have programmes to support home composting. More than twenty states have banned garden wastes from direct landfill disposal. In 1993, 18 pilot programmes were underway to evaluate composting of compostable (non-garden) wastes. These were mainly windrow systems with capacities up to 100 tonnes per day. Most took source-separated material from residential, commercial or industrial sectors.

In Japan, composting plants have been built in many cities, to reduce the waste load to incineration facilities. By the mid-1990s some 29 composting units were operating in Japan. In less prosperous nations, there is often a higher level of organic material in the waste stream. Brazil, which produces some 90 000 tonnes MSW each day with an organic content of up to 60%, now has 74 composting plants, 19% of which are fully mechanised. Around 30 municipalities in Brazil employ some form of source-separation for organic waste, further improving the quality of the final product.

The demand for compost made from source-separated materials should increase as experience in its manufacture and use grows. It is already used for spreading on areas of salty land to restore fertility, and as the world's topsoil is already being lost at the rate of three tonnes per acre every year it seems likely that compost will eventually be in demand for restoring the structure of farm soils everywhere.

A report from the European Commission[15] identified a number of effective centralised and home composting schemes in Spain, France, Ireland, Portugal, Italy and the UK. A major influence in the development of these schemes was the European Union's 1999 landfill directive, which sets high standards for waste disposal. This instrument includes provisions to reduce landfilling of biodegradable wastes to reduce environmental damage from the release of landfill gases and leachate. Some Member States have already introduced limits for the amount of biodegradable waste permitted to go to landfill.

[15] European Commission, *Success stories on composting and separate collection, ISBN 92 828 9295 6,* Catalogue No. KH-27-00-726, Environment DG Information Centre, BU-9 0/11, 200 rue de la Loi, B-1049 Brussels, Belgium, 2001.

The Commission's study concluded that successful diversion of biodegradable waste from landfill depends on separation at source. These wastes include kitchen waste (*e.g.* fruit and vegetable peelings), garden waste and (often) card and newspaper. Most successful schemes provide free containers of bags for collection. Frequency of collection varies from daily to once a week or alternate weeks. Organic waste collection must be frequent enough to prevent waste accumulating to levels unacceptable to the householder, especially in warmer climates.

The existence of end markets is extremely important; revenue from sales of compost or soil conditioners can fund the scheme. Use of the end product is fundamental in achieving the full environmental benefit of composting.

Many schemes share collection vehicles with schemes to collect dry recyclables. The scheme in Niort, France for example, achieved cost savings where the unit cost is half that of landfilling – additionally, a state landfill tax is avoided for all materials composted. Cost avoidance is an important benefit of composting. Most of the schemes surveyed in the Commission study had received additional financial assistance from local or national governments.

Good publicity and information was the most important overriding factor. Administering the schemes involved extensively the local municipalities or governments. Detailed planning and design are important when developing collection systems and treatment plants. Key success factors in the programmes were:

• clear achievable objectives
• right mix of waste types to targets
• effective and convenient to householders
• establishing a market for end product
• good financial management and planning
• good publicity and information

With the growing international trend of introducing a policy to ban landfilling of untreated organic wastes, composting is set to benefit from an increased enthusiasm for landfill diversion.

Thermal Treatment

Household waste has up to half the energy potential of coal. Recovering energy from the non-recyclable portion of household waste makes economic and environmental sense. The 230 million tonnes of municipal solid waste (MSW) created in Europe each year could meet 5% of Europe's energy needs. The main virtues of energy recovery are:

• waste volume reduction
• rendering waste inert
• recovering value from waste
• biodegradable waste diversion
• a practical method to manage increased waste arisings

There are four main reasons for public hostility to Energy from Waste (EfW) projects:

- reluctance to host new large-scale developments
- fears about the health impacts of emissions
- concern that EfW erodes enthusiasm to recycle, while still producing ash which must be landfilled
- anxiety of the scale of the plant – larger units imply a wider waste catchment area which can affect local networks

However, there are many examples of EfW plants blending in, usually because they have been there for many years and the communities are familiar with them. There are also examples of new facilities which quickly become part of the community. The unit in Dundee, Scotland signed a *Good Neighbour Charter*, which commits incinerator company Dundee Energy Recycling to standards above those required by law towards their neighbours in the community.

Dioxins. Dioxins are a family of chemicals, some of which are carcinogenic, which are present in the environment. Dioxins are also formed during manufacturing processes such as that for paper and in the metallurgical industries, as well as when organic materials such as coal and wood are burned.

A national study by the Swiss Environment Agency concluded that the largest source of dioxins in Switzerland is now the burning of household waste at home or in the garden. Controlled incineration of MSW results in the annual emission of 16 g dioxins, while domestic rubbish burning emits 27–30 g pa.

A materials flow analysis of dioxins in Denmark[16] suggested that total formation of dioxins between 1989 and 1999 is around 90–830 g pa. The most important contributing activity is waste treatment and disposal, and municipal waste incineration is the dominant source for dioxin generation.

The National Society for Clean Air & Environmental Protection[17] cites a survey of UK emissions of air pollutants which yields the data shown in Table 5. This suggests that MSW incineration contributed some 4% of UK dioxin emissions in 1998, a decline from 60% in 1990. The main factor in this fall was the closure of many older technology waste incinerators as a result of the 1989 EU municipal waste incineration directive.

Material versus energy. There is concern over whether *materials recycling* and *energy recovery* should be considered equally valuable. Most people agree that materials recycling should generally be considered before energy recovery, but that energy recovery is a valuable option within an integrated approach to resource and waste management. Because many recyclable materials, such as glass, aluminium and steel, are not combustible, their removal improves EfW plant efficiencies. Removal of wet organic wastes for composting also helps plant operation. Proper analyses of waste streams and volumes prior to plant construction can reduce conflict between material and energy recycling.

Ash management. Incinerator ash can be a source of pollution, but there is

[16] E. Hansen, *Substance flow analysis for dioxins in Denmark*, Environmental project 570 for the Danish EPA, 2000.
[17] P. Coleman, *Dioxin measurement, Clean air & environmental protection Journal of the NSCA*, Spring 2001, pp. 18–24.

Table 5 Sources of dioxins (1990 and 1998)

Source	1990 (g I-TEQ pa)	1998 (g I-TEQ pa)	(%)
Power stations (coal and oil)	35	18	6
Coal combustion (industrial + domestic)	38	17	5
Wood combustion (industrial + domestic)	26	26	8
Coke production	3	1	0
Sinter plant	42	43	13
Electric arc furnaces (iron and steel)	12	8	3
Non-ferrous metal production	27	22	7
Chemical industry	12	14	4
MSW incineration	602	14	4
Incineration – chemical waste	6	4	1
Incineration – clinical waste	51	24	7
Incineration – sewage sludge	5	3	1
Road transport	28	11	3
Accidental fires and open burning (farms)	121	64	20
Other sources	68	56	18
Total	1078	325	100

growing evidence that some of the material can be usefully recovered.[18] In America, there has been growing interest in marine applications such as inshore erosion prevention and artificial reef construction. In Germany, half the *incinerator bottom ash* (IBA, residual material from the furnace) is used as road base material and for sound barrier construction. The Netherlands aim to use 80% of all MSW incinerator residues – 40% of *incinerator fly ash* (materials trapped in incinerator air pollution control equipment) is used as aggregate in asphalt. Around 60% of IBA (more than two million tonnes pa) has been used in road base, embankments, and as an aggregate in concrete. In Denmark, bottom ash has been used since 1974. Almost three-quarters (72%) is used as sub-base in car parks, cycle ways and roads.

Energy Recovery Options

Landfill gas recovery.   Landfill gas is produced by the decomposition of organic wastes in the airless conditions of a landfill site. Landfill gas typically contains around 55% methane and 40% carbon dioxide together with small amounts of nitrogen, hydrogen and water. These gases can be collected using a network of horizontal pipes and wells, laid prior to and during filling of the site with waste. Beneficial use of landfill gas for energy evolved as a solution to the problem of potentially explosive gas leaking from landfill sites. Since methane is a greenhouse gas with a higher global warming potential than its combustion product carbon dioxide, using methane for energy recovery has the further benefit of reducing the net potential for global warming.

[18] Residua, *Ash handling from waste combustion: technical brief*, 1st Floor, The British School, Otley Street, Skipton, North Yorkshire BD23 1EP, UK, 1996.

24

Anaerobic digestion. Organic waste can be broken down using *anaerobic digestion* (AD), and the methane gas generated can be recovered. Anaerobic digestion has been used extensively for sewage sludge and for agricultural wastes. Its use to treat MSW, often with sewage sludge, provides a fuel which can be used, like landfill gas, to directly fire burners, to generate electricity or which can be cleaned and added to gas supplies. A major advantage of AD is that all the gas produced can be collected and used, unlike the gas in a landfill where collection efficiencies are relatively low (50% or less). AD also produces a solid residue or *digestate* which can be cured and used as a fertiliser.

Combustion. The conventional waste combustion technique is *mass burn incineration*, which involves the waste being burned as delivered, after removal of bulky items. To aid combustion, there is usually some mixing of the waste. In the past incineration plants were mainly designed purely to process waste, but today's plants are usually designed to recover energy (as steam, hot water or electricity) from the waste.

Refuse-derived fuel (RDF). The manufacture of refuse-derived fuel (RDF) is not new. It was originally devised as a means of avoiding the need to burn MSW immediately, and instead to turn it into a transportable, storable fuel. RDF production enables the subsequent thermal conversion of combustible portions of waste. While mass burning demands little sorting or processing of the waste, in RDF production the waste may undergo a number of pre-processing stages. At its simplest, RDF may be a coarse, fluff-like material produced from mixed MSW by a series of screening stages, plus magnetic removal of ferrous and non-ferrous metals. Alternatively additional processing may turn it into a densified, pelletised (or cubed) fuel, for ease of transport and storage. Turning waste into coarse or pelletised RDF differs from mass burn in being two-stage, where the first processing stage can be conducted completely separately from the second burning stage, and may be at a different site and time.

Fluidised bed combustion. Fluidised bed combustion technology is based on a system where, instead of the waste being burned on a grate (as in mass burn processes), the fire bed is composed of inert particles such as sand or ash.

When air is blown through the bed, the bed material behaves as a fluid. There are several different designs of fluidised bed (FB) combustors, for example *circulating* and *bubbling* beds. All need waste of uniform size. Compared to mass burn, fluidised bed combustion systems have reduced emissions, partly because of the process itself and partly because it is possible to add lime to the bed. Since as much as a third of the cost of mass burn plants is spent on the air pollution control (APC) system, there are savings to be made as fluidised bed systems have smaller APC needs. On the other hand, mass burn plants have no need for front-end processing of the waste. Also, as they are typically larger, and so benefit from economies of scale, the cost per tonne of waste processing in the two systems may not be markedly different.

Because FB systems are typically smaller, they can be more appropriate for smaller communities. The need to pre-process waste prior to combustion in an

FB combustion plant, in order to reduce its size and make it uniform, provides an opportunity to maximise materials recycling. However, while metals can be separated from the waste when it is being shredded and reduced in size, for successful recycling of most materials they must be kept clean and this requires them to be separated at source, not mixed with other wastes. While FBs have been used in industrial applications for a number of years, and to burn wood chips and similar single-material fuels, their use for mixed waste is more recent. Mixed MSW is not an easy fuel to burn, because of its variability, and maintenance costs of FB combustors used for MSW are likely to be much higher than those for a single, predictable waste stream like wood chips.

USA. There were 122 EfW plants in America in 1999, an increase of three on the previous year. Overall, the proportion of waste incinerated declined from 9% in 1997 to 7.5% in 1998, largely as a result of increased recycling and composting. Data from the US Environmental Protection Agency show that in relative and absolute terms the most recent data indicate that both materials recycling and energy recovery have reached a peak, together.

Europe. Brussels-based ASSURRE profiled incineration in Europe (2000), identifying 304 incineration facilities in 18 European countries, 96% of which recover energy. The average unit capacity is 177 000 tonnes per year. Units vary in size from an average of 83 000 tonnes per year per site in Norway to 488 000 tonnes per site in the Netherlands. On an annual basis, Europe has 50.2 million tonnes of capacity to treat household and related waste. 49.6 terawatt hours (TW h) of energy are recovered from MSW each year. 70% of this is used for district heating and 30% for electricity generation.

Types of energy produced vary between countries, depending on optimum technology and local demand. The annual amount of energy generated from incineration is equivalent to the electricity demand of Switzerland. Per capita, energy recovered from incineration is highest in Denmark, Sweden and Switzerland. It is lowest in Spain, UK, Italy and Finland.

Treatment costs vary by country, ranging from €25–30 per tonne in Spain and Denmark to €160 per tonne in Germany.

Gasification and pyrolysis. *Gasification* is the process of reacting carbon with steam to produce hydrogen and carbon monoxide. Gasification converts a solid or liquid feedstock into gas by partial oxidation under the application of heat.

Pyrolysis is a complex series of reactions initiated when material is heated (to around 400–800 °C), in the absence of oxygen, to produce condensable and non-condensable vapour streams and solid residues. Heat breaks down the molecular structure of waste, yielding gas, liquid and a solid char, all of which can be used as fuels.

Both technologies have primarily been used for specific and generally single, unmixed waste streams such as tyres and plastics, or to process RDF. However, one German pyrolysis plant has been processing MSW since 1985.[19] German

---

[19] K. Strange, *Advanced thermal treatment techniques – an overview*, IBC conference, The Future of Waste Management and Minimisation, Regents College, London, September 21–22, 1999.

waste company Deutsche Babcock's plant was commissioned in Günzburg, Bavaria in 1983. Since 1985 the plant has been in permanent operation. The shredded waste is fed into a gas-fired rotary drum where it is pyrolysed at temperatures of 400–500 °C. The gas passes through a cyclone for the removal of coarse particulates and is then directly burnt in a post-combustion chamber at temperatures of around 1200 °C. Despite the above example, neither pyrolysis nor gasification is generally considered suitable for handling mixed, untreated MSW in large volumes at present.

The most popular application today is apparently to displace conventional MSW incineration, although Juniper has also identified specific niche opportunities for refineries, the paper and pulp industry and waste tyre disposal. However, the lion's share of the burgeoning market will be in the conversion of agricultural residues into a renewable energy resource, either by direct combustion for power generation, or through the creation of bio-oils (synthetic fuels for vehicles).

These advanced thermal treatment (ATT) technologies are very varied, with more than 60 different systems deemed technically and commercially interesting. Applying ATT systems to MSW streams will increase in future mainly because these technologies are perceived to represent cleaner, more socially acceptable waste management systems than conventional incineration with energy recovery. Gasification and pyrolytic process can convert waste into molecular building blocks, to generate new feedstock compounds for the petrochemical sector: waste tyres and plastics can be recovered for material re-use and energy recovery. However, Juniper believes that the economic climate presently means that most applications will focus on combustion and energy recovery from the products recovered.

The key drivers encouraging the adoption of ATT systems are:

- reduction in the volume of waste for final disposal
- rendering the waste for final disposal inert
- recovering value from the waste (usually as energy)
- pressure for more sustainable waste management systems
- the complementary nature of materials and energy recovery
- diversion of biodegradable materials from landfill
- chronic present or anticipated future shortage of landfill capacity
- economic instruments, such as landfill taxes and alternative energy subsidies

It seems clear that the pyrolysis and gasification business is staking a claim to territory occupied by the conventional EfW incineration sector. Nowadays. a new EfW plant can cost between US$30–100 million to build, with operating costs of US$50–100 per tonne. A key factor is the increased pressure to reprocess materials within the policy context of waste material flow management. This leads to increased waste collection and source separation, which also tends towards the isolation of high calorific value fraction (and separately, a smaller quantity of hazardous materials).

Landfill

Landfill is often regarded as the last resort waste management option, an 'out of sight, out of mind' solution. While this may be partly true, modern landfilling is an active treatment process applied to most solid wastes. An engineered landfill is designed to contain waste and its decomposition products until they are sufficiently stable and inert to present no significant risks to health or the environment. Other benefits, such as material and energy recovery or land reclamation, may also be derived from properly designed facilities.

Municipal solid wastes became a problem with increased urbanisation. Waste disposal became a priority, not only because of the nuisance of waste dumped in the streets, but because of very real health risks. Epidemics of yellow fever, cholera, smallpox and typhus were not unknown. For a time, cities disposed of their garbage in rivers or lakes. Others simply dumped garbage in open pits on unused land. In many of the world's poorer countries, conditions for waste disposal are still rudimentary.

Modern landfills. A common landfill classification system for reflects the type of waste each receives. There are landfills for *hazardous* wastes, *municipal* wastes and *inert* wastes. In practice, these are not exclusive definitions. Other variants include mono-fills, in which single waste types are allowed, and co-disposal sites, in which municipal and hazardous wastes may be combined.

Landfills can also be classed by the management strategy employed:

Total containment. Virtually all movement of water through the landfill is prevented. Total containment imposes a long-term responsibility for monitoring and supervision. This strategy is often used with hazardous wastes, less frequently for municipal solid waste.

Containment and collection of leachate. Leakage of water from the landfill is controlled (but not eliminated) by using a low-permeability liner beneath the wastes, and by collection, removal and external treatment of liquid decomposition product (leachate). Risks of leachate pollution depend on the extent to which the containment barrier integrity is maintained and on the efficiency of leachate management. This strategy demands expensive, active systems, and research is underway into accelerated leaching – the *flushing bio-reactor* concept – to speed stabilisation, from perhaps centuries to just a few decades.

Controlled contaminant release. In this approach, the base liner is made of natural, often local, materials. While sumps for collection and removal of leachate are sometimes provided, leachate levels are permitted to rise within the waste, permitting gradual migration through the liner into the ground. Naturally, this approach is not suited to every location and geological setting, so full risk and environmental impact assessment approaches should be carried out before development.

Unrestricted contaminant release. Here, no control is used for water infiltration or leachate escape. This occurs, by default, in waste dumps, particularly in poorer countries.

Landfill Design and Operation

Before constructing a landfill, detailed preparations are required by licensing authorities. Landfills consist of areas (*cells*) of waste, spread and compacted in discrete areas. Compaction is often carried out, using bulldozers in poorer countries, and specialised steel-wheeled compactors elsewhere. These can achieve final waste densities of more than one tonne per cubic metre (t m^{-3}), although 0.7–0.8 t m^{-3} is more typical. At the end of each day, or more frequently, the waste is buried with a layer of cover material (usually soil) which is itself compacted. This controls nuisances, such as pests, litter and smells, and reduces the likelihood of fire.

The principal components of containment are usually the liner, its protection layer, a leachate drainage layer and a top cover. The most common liner materials are *mineral liners* (using clay, for example), *polymeric flexible liners* (of such materials as high density polyethylene) or *composite liners* using both approaches. Liners are more robust than the name might suggest; composite liners can be as much as five metres in depth.

Sanitary landfilling demands the isolation of wastes from the environment until they are rendered harmless through biological, chemical and physical processes of nature. Many complex reactions can occur between the extremely heterogeneous components of landfilled waste.

Biological processes. Generally, more than half of household waste is organic. This degrades gradually through five stages within a landfill: *aerobic hydrolysis*, in which micro-organisms convert some carbohydrates to simple sugars (such as glucose), carbon dioxide (CO_2) and water; *hydrolysis and fermentation*, when carbohydrates, lipids and proteins are broken down and fermented yielding volatile acids, acetate, CO_2, hydrogen (H_2) and inorganic salts; *acetogenesis*, where bacteria turn soluble acids to CO_2 and H_2. These, with carbohydrates are also transformed into acetic acid; *methanogenesis*, in which bacteria convert acetic acid to methane and CO_2. Finally, conditions may become *aerobic* again as the landfill becomes more stabilised.

Chemical processes. Two general types of chemical reactions take place within landfilled waste. Firstly, *oxidation*, using trapped oxygen, which soon becomes depleted. Secondly, *acid–metal reactions*, due to the presence of organic acids and CO_2. These processes mobilise metallic ions and salts which are potential pollutants. However, once methane generation is established the landfill becomes less acid and metals (especially mercury, lead and cadmium) are generally retained (as relatively immobile sulphides).

Physical processes. *Compaction* of waste has a strong bearing on its behaviour. This process begins at the collection stage, takes place at the landfill site and continues within the landfill itself. The effects of water as it passes through the landfill also have a profound influence on the long-term behaviour of the waste, dissolving soluble materials and transporting unreacted matter. The *absorption* of dissolved pollutants (by, for example, cellulose-based matter within a landfill), helps retain materials, at least before saturation occurs. Finally, *adsorption* is an important factor within a landfill, as wastes become bonded to the surface of other materials.

Landraising. Recent years have seen a growth in landraising, or above ground landfills. Although more visually intrusive, landraising does offer some environmental benefits. Wastes are kept further from any potential contact with groundwater, and leakages are easier to identify and control.

Pollution from landfills. A waste disposal facility must guarantee adequate control over the two main types of pollution – *leachate* and *landfill gas*. When landfill gas is collected some liquid (condensate) is also collected.

Leachate is generated as a result of moisture entry into a landfill, either as rain, snow melt, run-on or as moisture in the waste itself. A typical landfill design includes run-on control and a final cover to minimise moisture flow into the waste. The two most significant components of leachate are organic chemicals and heavy metals. These may be conserved in the landfill in the short term, but through biochemical processes can be mobilised and released. *Organic chemicals* are present as soluble decomposition products (*e.g.* organic acids). They are also present as organic chemicals (*e.g.* benzene, toluene, dioxins, halogenated aliphatics, pesticides, PCBs and organophosphates) discarded in the waste. *Heavy metals*, such as mercury, chromium, nickel, lead, cadmium, copper and zinc, are often found in landfill leachate. Discharges depend mainly on the acidity and the rates of flow of leachate. Many heavy metals come from the non-regulated hazardous waste fractions from households and businesses.

No engineering design can guarantee total containment, and some migration of leachate to contaminate groundwater supplies may occur. Policy-makers generally demand quality control systems and the use of comprehensive environmental impact and risk analysis techniques. Leachate management systems tend to be one of the three following types: on-site treatment (generally some form of aeration tank system), disposal to sewerage systems, or transport off-site for treatment elsewhere.

Landfill Gas

All landfills containing biodegradable materials will produce landfill gas. Typically this gas contains methane and carbon dioxide as major components in a wide range and combination of concentrations. The composition of gas within the waste changes quickly, with significant quantities of methane taking 3–12 months to be generated. Hydrogen and hydrogen sulfide may be produced at first, although degradation is initially aerobic – yielding carbon dioxide and water. Later, the process becomes anaerobic. Landfill gas emissions have a number of pollutants of concern to human health, such as acrylonitrile, benzene and carbon tetrachloride. Methylmercury has also been measured at the working face of landfills. Landfill gas emissions contain volatile organic compounds that contribute to urban smog.

Landfills are one of the largest man-made sources of methane, a potent greenhouse gas (GHG). Atmospheric methane levels have increased since the beginning of the 19th century. World-wide, emissions from landfills and open dumps have been estimated to contribute 6% of total global methane emissions. Landfill gas contains perhaps 55% methane, 45% CO_2 and more than 100 trace elements. In theory, one tonne of MSW will produce up to around 375 cubic metres (m^3) of landfill gas, with a calorific value of up to 20 MJ^{-3}. However, collection is challenging, and even the most efficient landfill gas recovery systems capture no more than 70%. As a general rule, a landfill containing one million tonnes (Mt) MSW disposed over ten years will generate a peak of 700 m^3 h^{-1} methane at peak.

It has been estimated that annual global production of methane from solid wastes will rise from around 55 million tonnes (Mt) in 1995 to 90 Mt in 2025. In 1998, the European Commission[20] published a study which concluded that in 1994 around 22 Mt of methane were emitted from man-made sources in Europe, including 8.2 Mt from landfills.

Venting systems are used to prevent landfill gas reaching dangerous levels. These systems range from something simple like a gravel seam, which allows gas to flow to a particular zone, to sophisticated networks of vertical boreholes and horizontal inter-connectors. Full recovery schemes include collection, extraction and transportation elements, and are often installed progressively, as the landfill is constructed. Some landfill gas is sure to escape, partly because methane is less dense than air. Where the landfill is covered by an impermeable cap, generated gas will tend to move laterally beyond site boundaries, especially within deeper sites and where gas migration control systems are not in place.

Methane can be recovered and used, which can reduce costs. By the early 1990s, as many as 500 landfill gas recovery projects existed world-wide, usually generating electricity. Most schemes are in America, where there are more than 200 projects, although Britain, Germany and Scandinavia make great use of this resource. Canadian landfills annually generate one Mt of methane per year, equivalent to nine million barrels of oil. This would meet the annual heating needs of 500 000 Canadian homes. More than a quarter of this is captured at 27 landfill gas recovery schemes. In Germany, almost two-thirds of municipal waste landfills recover energy from landfill gas.

Reclamation

When the landfill is complete, the surface is sealed; the cells are capped with an impermeable membrane, and covered with more than half a metre of soil (ideally, this should be the original top-soil, collected and stored at the formation of the site). The final cover minimises the passage of water into the landfill, reducing leachate flow and the migration of gases. It also provides a further barrier between the waste and the environment, while allowing plant growth and

[20] European Commission, *Options to reduce methane emissions* (*Final Report*), a report produced for DGXI, November 1998.

landscaping. An increasing trend is for landfills to be converted into recreational areas such as parks and golf courses. A good example of this is the Danehy Park site at Cambridge, Massachusetts, USA. This has been open to the local community since 1990, and includes three softball fields, three soccer fields, and 50 acres of jogging, walking and cycle trails. Remaining gas travels laterally along a crushed stone passive venting trench, and only a small amount of settlement has taken place since the park opened.

Disincentives

Many governments have tried to of reduce dependency on landfilling. For example, the British Government introduced a Landfill Tax in 1996, which reached £12 per tonne for active waste and £2 per tonne for inert waste in 2002 (the active waste tax will rise to £15 per tonne by 2004). An enquiry suggested that there was no significant change in the amount of active waste disposed, although there was a 20–30 Mt reduction in the disposal of inert waste (mostly construction and demolition material).

Landfill Bans

Many countries ban certain materials, if untreated, from landfills. A US survey showed that virtually every state (or local authority) now operates bans, though not all are enforced. In California, bans cover latex paint, white goods, automobiles, recyclable metals, lead–acid batteries, adhesives, automotive products (*e.g.* anti-freeze, transmission fluid), cleaners, pesticides, mercury, solvents, used oil, whole tyres and household batteries.

A key policy development which builds on the concept of the landfill ban has been the European Union Landfill Directive (1999/31/EC). The main provision of this directive is the progressive banning of municipal biodegradable waste from landfills, to 35% of 1995 levels by 2020. In the UK this means at least 6 Mt pa must be diverted (this could reach 33 Mt pa if arisings continue to grow). Currently, more than 80% of MSW in Britain is landfilled, which means that more than 60 composting facilities, up to 120 materials recovery facilities and perhaps 50 energy from waste plants will be needed.

Some countries in the EU, such as Denmark, have traditionally depended less on landfill, and already comply with the directive's targets. In America, landfills currently manage 55% of MSW generated (120 Mt pa). There are now fewer municipal solid waste landfills in the US than in the 1980s, but the average size has increased. As recovery rates have increased and combustion has remained constant, the percentage of MSW discarded has steadily fallen.

Some municipalities in Australia and Canada have advanced the concept of regarding landfills as long-term stores of material, for future use when economic changes have transformed a waste with no value to a resource. If long-term environmental monitoring is required at such a facility, then perhaps this is no more sustainable than the alternatives.

Landfills are unwelcome as neighbours, and regularly attract a hostile response from prospective host communities, yet these facilities will continue to be necessary. However much we endeavour to reduce wastes, to increase re-use and recycling, to compost and to recover energy, the laws of nature mean that there will be some residual matter for which society can find no further use.

5 Recycling

It is widely recognised that waste management practices that continue to favour landfill over other waste management options are unsustainable.[21] New, integrated, systems are required to promote the more efficient use of resources and the parallel economic, social and environmental objectives of sustainable development. This has prompted a series of measures at both national and European levels to divert wastes to other management routes. However, as in shown in subsequent section, it has proven difficult to increase recovery through material recycling for many components of MSW.

There are a number of reasons why a higher level of recycling does not occur. These can be divided primarily into:

- *market failure*, where prices do not reflect environmental resource values
- *government failure*, where policies put in place by governments may encourage inefficient practices
- *institutional failure*, where public awareness may discourage socially optimal strategies.

Also, the cyclical nature of markets has led to depressed prices and even the collapse of voluntary schemes, as well as reduced profit margins for the private sector, which passes on its costs to local authorities. This has discouraged recycling schemes.

ERM[21] cites further barriers to increased levels of recycling by local authorities:

- national distribution of secondary materials dealers and reprocessors
- limited quantities of recyclate arisings in each authority and the low resource density in rural areas
- low demand caused by perceived and actual performance of recyclate in respect of material specifications
- contamination of separated materials
- duration of collection and disposal contracts
- limited public participation, often associated with distance from recycling centres

The most common strategy for improving recycling rates is to improve the separate collection system, by increasing the coverage of materials collected at the kerbside and by increasing the convenience and density of recycling centres. Table 6 shows how other countries plan to increase recycling.[21]

Waste Watch[22] has assessed where money could best be spent to increase recycling in Britain. Waste Watch concludes that Waste Disposal Authorities (WDAs, responsible for civic amenity sites and for treatment of compostables)

---

[21] ERM, *Research study on international recycling experience,* for the Department of Environment, Transport and the Regions, May 2000.

[22] Waste Watch, *No waste of money – how recycling can be funded,* ISBN 1 898 026 28 9, Waste Watch, 96 Tooley Street, London SE1 2TH, UK, 2002.

Table 6 Future developments in recycling programmes assessed

Case study	Market development	Increased collection	Increased information	Fiscal incentives
Denmark		X	X	X
France		X		
Germany		X	X	
Italy		X	X	X
California	X			
New York		X		
Seattle	X	X	X	
Ontario		X		
Canberra	X	X		
New Zealand	X	X		X

could make significant impacts on their targets by re-organising their civic amenity (CA) sites, and/or investing in depots or separation facilities. This would extract value from materials which would otherwise have to be sent to landfill. Waste Collection Authorities (WCAs) can meet their targets by increasing the range and volume of materials collected through bring schemes or kerbside recycling schemes. WCAs could also increase their recycling rates by reducing the amount of waste collected in the first place. They may do this through encouraging waste minimisation (through active minimisation, promotion of reuse/refurbishment, *etc.*) or through reducing the frequency of collection or collection of certain types of wastes such as trade waste. The choice of approach will determine both the level of performance and the overall costs.

The choice of approach will also be determined by the target rate. It is unlikely that WCAs with targets of 30% and above could be sure to meet these figures without collecting organic wastes. Organic waste collection does not necessarily cost more than a dry recyclables collection. However, the collection of organic waste will require new approaches to treatment.

6 What Is Integrated Waste Management?

There are many different ways of dealing with waste in order to minimise risks to public health and the environment. For many years waste management was carried out in a piecemeal, relatively unplanned way. More recently, experience has shown that a more sustainable approach to society's uses of resources and management of wastes is needed.

The term integrated waste management (IWM) is often used to describe an approach in which decisions on waste policies and practices take account of waste streams, collection treatment and disposal methods, environmental benefits, economic optimisation and social acceptability.

Integrated Waste Management systems combine waste streams, waste collection, treatment and disposal methods, with the objective of achieving environmental benefits, economic optimisation and societal acceptability.

Principles of IWM

IWM allows decisions to be based on best practice and cost transparency. The smaller the amount of waste put into the system the lower the costs apportioned to the generator of that waste. This provides incentives for users to reduce the amount of waste they generate.

- IWM considers all options (collection, recycling, composting, biogasification, energy recovery and landfilling) for the entire municipal solid waste stream – not simply sub-streams
- shared responsibility – manufacturers, distributors, retailers, consumers and other stakeholders have a responsibility to support IWM. Each group is responsible for the correct management of wastes they create
- three criteria should be considered – environmental effectiveness, economic efficiency, social acceptability
- flexible application for different communities and regions
- transparent costs for waste management
- market-oriented recovery and recycling
- appropriate economies of scale
- continual assessment to accommodate changes in quantity and quality of the waste stream

IWM is a concept which has different local applications and which depends on many variables such as the composition of the waste stream, infrastructure, markets for recyclables, budgets, local legislation and land availability. IWM seeks the best options for waste management, with an emphasis on evaluating all available strategies to deliver more sustainable systems.

There are several possible aspects to this integration. Upstream collection and handling should be integrated with downstream treatment, processing and disposal. The management of municipal solid waste (MSW) may be integrated with industrial and commercial waste streams. The collection of materials for recycling or composting needs to be integrated with markets for products for the system to be sustainable. A further tier of integration rests at the policy level, enabling the development of new waste management facilities. This calls for collaboration between planning authorities, industry and the public.

The details included in the IWM concept vary around the world. In California, for example, IWM specifically includes public education and outreach programmes, along with efforts to foster markets for recyclables. In South Africa, IWM aims to integrate and optimise waste management in order to maximise efficiency and minimise environmental impacts and financial costs of waste and improve the quality of life of its citizens.

The term IWM does not describe or prescribe the actual techniques applied to deal with a particular waste stream, but refers to the overall approach to considering how best to manage wastes. An integrated approach does not mean that a community or region would implement all waste management techniques, only that they are all considered.

The variable nature of waste means that no single option (apart from landfilling, and there are sound reasons for wishing to reduce dependence on final disposal) could ever deal with an entire waste stream. Some waste is suitable for recycling, some for biological treatment, some for energy recovery. Policies can provide a framework where more or less waste might be channelled in particular directions, but there are limits, beyond which each option becomes technically unreasonable, environmentally adverse, economically inefficient and socially unacceptable. The upper limits to recycling are seldom higher than 60%, while the organic content of waste (which limits scope for biological treatment) rarely exceeds 70%. The proportion of municipal solid waste which is combustible is often less than 60%.

Operations within any waste management system are clearly connected. For example, the type of sorting scheme introduced in an area will affect the ability to recover materials, or produce marketable compost. Many effective recycling programmes deliver a residual waste which becomes more suitable for energy recovery. Landfilled waste can be reclaimed, and landfill gas recovered.

An integrated system would include an optimised waste collection system and efficient sorting, followed by one or more of the following options:

- *materials recycling* will require access to reprocessing facilities
- *biological treatment* of organic materials will ideally produce marketable compost and also reduce volumes for disposal. Anaerobic digestion produces methane that can be burned to release energy
- *thermal treatment* [such as incineration with energy recovery, burning of Refuse-Derived Fuel (RDF) and burning of Paper and Plastic-Derived Fuel (PPDF)] will reduce volume, render residues inert and should include energy recovery
- *landfill* can increase amenity *via* land reclamation but a well-engineered site will at least minimise pollution and loss of amenity

To manage all solid waste arisings in an environmentally effective way requires a range of the above treatment options.

Who Pays?

Inevitably, an issue at the heart of waste management decisions is who should pay. In general, industry tends to believe that the owner of waste at the point of disposal should pay for its management. Often, policy-makers prefer that the costs are imposed on the first producers, importers or retailers (this cost is of course passed back to the consumers, who will also have to pay their taxes for disposal of their residual waste). This was the case in European legislation on waste electrical and electric equipment and on end of life vehicles. In draft legislation on these waste streams, the European Commission considered that the final user should not face any disposal costs which might discourage proper waste management.

Shared responsibility assigns costs of waste management to the owner of the product at each stage of its life cycle and provides a clear financial incentive for each actor in the life cycle to reduce the amount of waste they produce. This contrasts with *extended producer responsibility* which puts the costs of all environmental impacts throughout the life cycle of a product from production, through consumption, disposal and recovery on the manufacturer, who, as mentioned above, has to incorporate this cost into the price of the product.

Local authorities play an important role in providing (or contracting for the provision of) services to collect and manage municipal solid waste. In these cases, householders pay directly for the services. Municipal taxes and charges are the usual methods of payment for existing waste management systems. However, these are often set at a fixed level, again providing little incentive for householders to minimise waste.

Local authorities are increasingly charging householders in proportion to the quantity of waste they produce – commonly known as variable rate charging. Variable rate charging does lead to increased participation in recycling/composting programmes, or to consumers actively seeking products that generate less waste. In Oostzaan (Netherlands) the introduction of variable rate charging in 1998 led to a 38% reduction in waste arisings, and a 60% fall in the generation of residual waste (waste neither recovered nor recycled).

It became clear early in 2002[23] that the Government of the Republic of Ireland is pursuing this policy and other economic instruments with enthusiasm. Householders in Ireland are to be charged by weight for the waste they put out for collection within three years under a new policy aimed at providing more incentives for re-use and recycling.

The Minister for the Environment, Mr Dempsey, also announced that a landfill levy will be imposed on all local authorities and private contractors from June 1, 2002 – initially at €15 per tonne, increasing by EUR5 per year. National bans on landfilling specific recyclable materials are also to be introduced this year to support greater recovery rates.

There are various methods of variable rate charging associated with IWM programmes around the world, based on different charging systems:

* bin size and frequency of collection (Copenhagen, Denmark)
* bin size (Seattle, USA)
* frequency of collection and household type (Helsinki, Finland)
* property value (Hampshire, UK)
* apartment size (Brescia, Italy)
* per inhabitant charge (Lahn-Dill Kreis, Germany)
* number of rooms and waste bag charge (Zurich, Switzerland)

Variable rate charging schemes can reduce the amount of household waste generated and also reduce the actual per capita cost for the service, while increasing recycling rates. In Hainaut province, Belgium, increased landfill costs and high waste collection taxes prompted the authorities to raise additional monies *via* variable rate charging. Unit pricing (per bag put out for collection) resulted in 40% less waste going to landfill. In some areas those who exceeded a certain tonnage disposed to landfill were forced to pay a penalty. Selective collection for recycling doubled, as did the usage of the collection container areas.

Similar results were recorded in the North America, where there are several

---

[23] *Irish Times*, March 13, 2002.

thousand *pay as you throw* schemes in operation. In the US state of Maine, per capita household waste collected for disposal was 55% less in municipalities operating a variable rate system, and per capita costs dropped by 25%. Variable rate systems led to more kerbside recycling and increased the amount of compostables diverted from landfill, contributing on average 8–11 percentage points of diversion from landfills.

There is growing support for the basic concept of IWM. The concept is supported by the UK Environment Agency (WISARD – Waste – Integrated Systems Assessment for Recovery and Disposal) and the US Environmental Protection Agency who have both invested in the development of computer models to optimise integrated systems. The concept is also addressed and supported in the UK Waste Strategy 2000.

There is a growing body of evidence that IWM is being increasingly adopted at a local level, resulting in increasing levels of recycling and energy recovery and reducing the current dependence on landfill. Although there is still not full consensus about the specific details of integrated waste management systems, this is to be expected as IWM is for local application, so systems will differ. Neither is it certain whether the majority of legislative bodies are yet in favour of a process which, by its nature tends to encourage a more flexible approach to managing wastes. Flexibility can make the introduction of standards and regulations more difficult. However, there is little argument that flexible, responsive and planned systems are more likely to deliver social, economic and environmental benefits than pursuing a dogmatic approach which obliges pursuit of targets which may not be justifiable.

Dogma and debate over the relative priority accorded to recycling and energy recovery lost much of its intensity during the 1990s. There is now considerable agreement amongst proponents of energy recovery that all reasonable efforts should be made to recycle materials from waste, before turning to energy recovery. Conversely, many recyclers recognise that appropriately scaled and sited incinerators can complement materials recovery programmes.

IWM is as much a state of mind for policy-makers and other stakeholders as a practical approach to standardised problem-solving. Like sustainable development, IWM is a concept which is becoming increasingly used (and sometimes mis-appropriated) and which describes a methodology as much as an end-point in a process.

Waste management systems evolved from basic practices designed to protect human health and are now becoming more sophisticated and geared towards environmental protection. With the drive in recent years towards sustainable development, the need to seek environmental, economic and social perspectives to waste management has led to the wider use of IWM. The next step is to integrate more fully the management of resources and waste to achieve truly sustainable waste management systems.

In order to plan IWM systems that are environmentally and economically sustainable, their economic costs must be established. In addition, the following factors need to be addressed:

- good system management
- vision – clear and defined long-term strategies
- stability
- economy of scale or mass balance
- landfill space
- control of all municipal solid waste arisings
- enabling legislation
- availability of funding
- public opinion

7 Trends In Waste Management

The European Environment Agency (EEA) predicts that per capita consumption in the EU is expected to continue to increase up to 2010. The baseline scenario used (based on OECD and EC socio-economic business-as-usual scenarios) projects a 45% economic growth between 1990 and 2010, and 50% increase in final consumption between 1995 and 2010. Notwithstanding the current limitations of waste data, the EEA has made predictions of total waste generation. Household waste generation in the EU is estimated to grow by around 20% to 2010 (ETCW, 1999) based on forecast increases in per capita consumption over the period 1995–2010.

The main challenge therefore will be to break the linkage between consumption and waste generation, if sustainability in waste management is to be achieved. Unless the linkage can be broken through strict implementation of a range of policy initiatives, we can expect to see a continuing trend of increasing waste generation throughout Europe. Increasingly this is being recognised in national recycling plans, with Denmark being a notable example.

The EEA found that landfilling remains the most common municipal waste management option in Europe. As a proportion of total waste handled in the countries of the European Union, ETCW believes it was actually higher in 1995 at 67% than over the period 1985–1990 at 64%. Over this period the absolute amounts of MSW landfilled increased from 86 to 104 Mt pa.

In a study carried out by Enviros for the Resource Recovery Forum (2000), Austria, Denmark, Germany, Netherlands, Sweden and Switzerland have all reduced MSW disposal to landfill during the 1990s. Only France, Norway, and Spain have increased volumes to landfill.

The growth in recycling and composting for four countries exceeded the underlying growth in municipal waste arisings over the various periods in question. These countries were Austria, Netherlands, Sweden and Germany. Germany however, showed a significant growth in recycling and composting, whilst appearing to decrease the total volume managed by three million tonnes (with a somewhat limited dataset).

In Denmark, enhanced recycling offset the growth in waste arisings and the two are now broadly in equilibrium. Similarly, Switzerland's recycling and composting performance was similar to the overall growth in waste managed. Three countries have MSW growth levels that in absolute terms outstripped the quantities of waste recycled and composted over the same period. These countries were France, Norway and Spain.

Those countries that have successfully countered waste growth through enhanced recycling and composting have benefited from a rapid development of the required infrastructure complemented by a parallel growth in energy from waste (EfW) capacity. Progress in increasing recycling and composting in many of these countries has been supported by the introduction of a number of economic and policy measures.

In all these countries there was a fundamental shift in waste policy in the early 1990s, moving away from a focus on collection and disposal towards an integrated set of measures, which have included the development of a planned infrastructure of recycling, centralised composting schemes, and energy from waste incineration. Most of these countries have already introduced limits or bans to landfill. This group of countries is also characterised by high waste treatment costs, and a high level of EfW capacity per capita compared to the UK.

Notwithstanding any efforts at waste prevention, a need exists to develop further complementary forms of downstream waste treatment. The response of many European countries to the Community Strategy and the EU landfill directive in particular has included a substantial increase in recycling, composting and EfW capacity.

At the national level, significant progress in reducing landfill and increasing recycling and composting over and above the underlying growth in MSW has been made in Austria, Denmark, Germany, the Netherlands and Sweden. Notwithstanding the range of regulatory and policy instruments in place, these countries have benefited from rapid development in recycling, composting, anaerobic digestion and EfW infrastructure.

Data from Eurostat (2001) echoed these findings, while noting that some parameters are still not defined in a harmonised way, making comparisons at a European level problematic. Eurostat summarised the findings as follows.

Denmark

In 1994, 2.75 Mt of MSW were collected – an increase of 34% in comparison with 1980 (2.0 Mt). Household waste accounted for 72% of the municipal waste collected in 1993 and 1994. 23% of the waste produced was recycled in 1994, in comparison with 17% in 1993. There was also an increase in landfill use over this period (from 20% to 22% of waste). The amount of waste incinerated remained stable in 1993 and 1994 (1.5 Mt) and incineration dealt with 63% of MSW in 1993, and 56% in 1994.

Spain

The amounts of municipal waste collected increased by 67% between 1989 and 1998 (from 12.5 to 21 Mt). The average production of waste per inhabitant also increased significantly, by 66% between 1989 and 1998 (from 322 kg to 533 kg).

In 1998, household waste accounted for an average of 90% of municipal waste, with the one exception of the Balearic Islands where the figure was only 48%. The remaining proportion was formed by waste from small enterprises. In 1998, 76% of municipal waste collected were landfilled, the same percentage as in 1989 (77%) and representing an increase of 63% in amounts over this period.

France

The total amount of MSW collected increased by 12% between 1993 and 1996 (from 33.5 to 37.6 Mt). Household waste accounts for 60% of municipal waste.

Ireland

Ireland has experienced a 50% increase in municipal waste collection between 1993 and 1998 (from 1.3 to 2.0 Mt). The production of waste per inhabitant increased by 22% during this period, from 426 kg per inhabitant in 1993 to 613 kg in 1998. In 1985, all municipal waste was landfilled. In 1993, 9% of this waste was recycled. However, the amounts of waste landfilled doubled in practice as a result of the increase in the total amounts generated. There is no incineration of municipal waste, although a facility is planned.

Italy

The amount of municipal waste collected increased by approximately one third between 1991 and 1998 (from 20.0 to 26.8 Mt). This increase took place solely between 1991 and 1993. After 1993, amounts levelled out in some regions and in some cases even fell.

Landfilling is the most widely used method of waste disposal in Italy, accounting for approximately 77% of waste in 1998 – a slight reduction in comparison with 1996 (83%). In 1998, 7% of municipal waste was incinerated and this method was in use in eleven of the 20 Italian regions (significantly in four). In 1997, other forms of waste elimination or treatment, especially composting, accounted for 10% of waste (5% in 1996). The number of incinerators increased by 11%. In 1997, most of the 38 incinerators were located in the north of Italy. The number of composting sites or other treatments more or less doubled between 1996 and 1997 (from 60 to 115).

Netherlands

The amount of municipal waste collected increased by 6% between 1993 and 1997 in the Netherlands. The amounts of household waste increased by 66% between 1985 (5.2 Mt) and 1997 (7.9 Mt) accounting for 81% of the municipal waste collected in 1997. On average, the production of waste per inhabitant increased slightly by 4% between 1993 (595 kg) and 1997 (618 kg).

In 1995, 2.9 Mt were incinerated. In 1999, this increased to 4.9 Mt, because some large incineration plants were commissioned. The number of incineration plants did not change between 1980 and 1999, but the incineration capacity doubled. The number of landfills in use fell eight-fold between 1985 and 1997 (from 373 to 44 landfills).

Austria

MSW collected increased by 29% between 1989 and 1998 (from 2.5 to 3.2 Mt). This increase took place largely between 1993 and 1996. Subsequently, amounts seem to have begun to decrease. In 1998, household waste accounted for 100% of the waste collected in all regions except Tirol (59%), Burgenland (68%) and Wien (85%). In 1989, at national level, the percentage was 86%, falling to 59% in 1993.

The average production of waste per inhabitant increased by 22% between 1989 and 1998 (from 326 kg to 399 kg). In 1993, 55% of municipal waste was landfilled (1.37 Mt) whereas 75% of waste was landfilled in 1989 (1.83 Mt). The amounts of waste composted and recycled increased on average by one third between 1989 and 1993, accounting for 31% and then 44% of municipal waste.

In 1993, only Vienna used incineration (and used this approach for 66% of MSW, compared with only 20% in 1989). In 1995, incinerators were, however, to be found in all regions – with the exception of Salzburg.

Finland

MSW amounts collected in 1985 and 1997 were identical (some 2.5 Mt). The amounts varied, however, throughout this period, reaching a peak of 3.1 Mt in 1990. In 1994, household waste accounted for 41% of the municipal waste collected in comparison with 48% in 1985.

It would seem that the amount of waste per inhabitant fell by 7% between 1985 and 1997 (from 509 to 476 kg). In 1990, however, production per inhabitant rose sharply to 623 kg. Waste incineration, accounting for only 3% of waste in 1995, is very marginal in Finland. Of the two incinerators surveyed in 1990, only one was still to be found in 1995.

Summary of Trends

Across the European Union, municipal solid waste production fell nationally in only three countries: Germany, Finland and Sweden.

The regional situation is less clear-cut, since there was a reduction of municipal waste in approximately 40% of the regions studied, this percentage being influenced largely by the German regions. Municipal waste production in the poorest regions is catching up with production in regions with a higher GDP. In addition to a clear divide between the less prosperous southern regions and the northern regions, other features can be seen as regards waste production per inhabitant. In 1989 (or a relatively close year), production per inhabitant in the ten regions with the lowest GDP was lower than in the wealthiest regions, averaging 329 kg per inhabitant and 456 kg per inhabitant, respectively.

There was, nevertheless, a very substantial increase in municipal waste production in these regions, averaging 36%, which therefore closed the gap between the other wealthier regions in which the amounts generated increased by only 6%.

The percentage of municipal waste landfilled is increasing in Denmark, Spain and Portugal, but is decreasing in Germany, France, Ireland, Italy, Austria and Sweden. These trends relate to different reference years in different countries.

Trends in Incineration in Europe

There is a trend towards larger, more economic plants for incineration with energy recovery with better environmental performance, improved energy efficiency and lower unit operating costs. There is also a complementary trend towards small units adapted to local geography and the desire to contain transport costs. These reflect local conditions and greater public acceptance of small units.

The mid-size units (200–300 000 tonnes) will be less common in the future. There is pressure for tighter standards and greater public scrutiny in terms of protection of health and the environment. With the right investment in technology, all existing and new installations can meet EU standards as stipulated in the new EU Incineration Directive (2000/76).

The continued availability of cheap landfill remains the main obstacle to increasing the proportion of MSW from which energy is recovered. Economic instruments may be introduced to overcome this barrier.

As deregulation in the energy sector gather pace, the private sector will play a greater role in ownership and plant operation.

UK consultants Juniper[24] report that there are already more than one hundred pyrolysis or gasification facilities now either operating or on order. These are treating a number of waste streams, including MSW, agricultural wastes, sewage sludge and industrial wastes. Juniper have forecast that more than 200 new pyrolysis and gasification plants would be built between 1999 and 2010 – an investment potential of US$11 billion. By 2008, installed capacity may exceed 20 Mt pa, compared with 4 Mt pa today. The typical plant built to handle municipal solid waste (MSW) is likely to have an annual capacity of more than 100 000 tonnes.

8 Public Attitudes

Many of the technical issues surrounding waste management can be regarded largely as being understood and solved. It is in the area of our own attitudes that we must seek the most effective, enduring solutions to sustainable consumption. There is a substantial gap between perception and reality, with public surveys often showing that we have a high level of faith (or claimed faith) in our own environmental credentials, but little in that of our neighbours. In reality, there are very few people who can be considered truly green.

[24] Juniper, *Gasification of Biowastes Poised for Rapid Growth Predicts New Study* – Press Release issued by Juniper Consultancy Services Sheppards Mill, South Street, Uley, Gloucestershire GL11 5SP, October 24, 2001

Britons regard the UK as the worst country in Europe at protecting the environment, according to the latest *Yellow Pages Annual Recycling Survey*.[25] Sweden is ranked the greenest country in Europe, with Germany a close second. But the list of poor performers is easily topped by the UK, followed by Spain and France.

Households show a poor track record on recycling with 16% of people polled admitting to recycling nothing at all. This is up from 13% in the previous survey. Meanwhile, 31% say they recycle 'only a small amount' – down from 35%. Asked why they do not do more, a third of households who say they recycle nothing, or 'only a small amount', say they cannot be bothered. Three in ten say their nearest site is too far away, while 16% say they mean to do more, but just never get round to it.

Just over half – 53% – of UK households do, however, claim to recycle everything or 'quite a lot'. This is similar to the 52% recorded in the previous survey. Of those who do recycle, the latest survey shows an increase in the recycling of newspapers and magazines, glass, cardboard, plastics and garden produce.

The majority of people surveyed – eight out of ten – buy recycled goods though, with recycled toilet paper top of the poll. And 60% of all respondents are willing to pay more for goods with a perceived environmental benefit. When asked what was the most polluting item they use, one in five people single out their car, 39% of whom would be willing to swap it for something more environmental. Of course there is a difference between what people claim, and what they do.

This is not simply a British or European reaction. Recycling is not ranked very highly in the life of the average South African. The South African (SA) publication *Business Day*[26] reported that a mere 11% of adults in SA take the effort to recycle material such as paper, glass, plastic and vegetable waste. A sample of 2485 households was used, spanning all races and income groups and representing 15.9 million, or 91%, of all urban adult South Africans. According to the 11% of respondents who claimed to do any recycling, the most recycled material is newspaper, at 66%. Despite its weight, glass is the second most recycled waste material, with 40% of the recycling community separating it from their rubbish.

Other materials recycled by respondents include 'other types of paper' (by 36%), plastic supermarket bags (32%), drink cans (30%), 'other kinds of plastic' (25%), and tins (21%).

Only 2% of people said they did not recycle because they didn't know about it, so South Africans are at least aware that it takes place. Almost 40% of South Africans said they didn't recycle because they don't have enough material to make a real difference. However, even among recyclers, understanding of the options is limited. The main reason given by recyclers for not collecting glass, for example, was that they did not have enough to be worth collecting. But one in ten claimed not even to know that glass could be recycled.

A similar lack of awareness applied to other materials: the figure for

[25] *Yellow Pages, UK worst in Europe on the environment*, Annual Recycling Survey – news release March 14, 2002, YELL, publishers of Yellow Pages directories.
[26] Business Day, *Recycling seen as a waste of time for most households*, February 27, 2002, Business Day (Johannesburg), PO Box 1745, Saxonwold 2132, South Africa.

newspapers is 17%, 'other paper' 13% and plastic bags 12%. The lowest level of ignorance concerns cans, demonstrating the efficacy of Collect-a-Can's public awareness campaigns. Practicalities also have a strong influence on the level of recycling. A quarter of all urban adults said they did not recycle because they had nowhere to take the waste.

The result is that materials that have well-known collection points, such as glass and newspaper, have noticeably higher recycling rates. However, South Africans are clearly averse to going to too much effort in this regard. Only 34% of people said they would be prepared to take waste to a processing point.

Only half of the non-recycling community said they would co-operate in a recycling project if waste were collected from their homes, while a quarter said they would not recycle even if the material were collected in this way. Increasing the number of collection points would also not provide a complete solution, as 31% of South Africans said they would not be prepared to take waste to a collection point, even if it were no further than 500 m away.

The UK Resource Recovery Forum[27] has published a report exploring the view of householders in London, looking at attitudes to recycling and reasons for participating (and not participating) in local schemes.

Household Management

Most people feel that their lives are too busy and that they never have enough time to get everything done. People save time by sleeping less and compressing household chores, especially by shopping in supermarkets and cooking convenience foods. Even though many of them work outside the home, women generally set the rules for domestic management, including whether and how recycling is done. Because women continue to do most of the household chores they are, by default, the ones who throw most away. The exception to this pattern is household maintenance 'jobs', such as DIY and car servicing, which are still done largely by men (even in many single-female households). Some DIY products – paint especially – are thrown away intermittently, when garden sheds or cupboards are 'spring cleaned'. Gardening is done by both men and women.

Households are generally more concerned with untidiness ('mess') than waste ('rubbish'). Items are typically thrown away unconsciously as part of an overall effort to keep clutter at bay. Some materials are disposed of as they arise – typically food-related items including (storable) packaging – while others may accumulate to be disposed of once a week – newspapers and bathroom items, for instance.

Recycling does not fit easily into domestic routines, except where there is an easy to use kerbside collection scheme. People typically make a special effort – develop a new habit – if they want to recycle. Against a background of busy lives it is much easier for people to 'opt out' of recycling than 'opt in' – except where there is a very easy to use kerbside scheme, or unless they have a particularly strong pro-environmental commitment. Recycling is often initiated and sustained by women, as part of their overall control of household routines. Some keen

[27] RRF, *Household waste behaviour*, Resource Recovery Forum, 1st Floor, The British School, Otley Street, Skipton, North Yorkshire BD23 1EP, UK, 2004.

recyclers (including men) continuously have to remind other family members to separate items for recycling.

Given the ways in which households are managed – and work and domestic life are precariously balanced – people are likely to resist the call to do more recycling if they perceive it as 'extra work'.

Claimed Recycling Habits – Participation and Frequency

People typically over-claim participation in recycling schemes in questionnaire surveys. Checks in the London household survey suggest at least 10% over-claiming. Other research, which has compared claims with waste recovered, suggests much higher levels of over-claiming. Bearing in mind likely over-claiming, the household survey indicates that more than half of London households are doing little or no recycling at present. Up to another third of households are recycling but think they are doing as much as they can. Only around a fifth are recyclers who say (without prompting) that they do less recycling than they should.

(Claimed) high recyclers tend to be older and more affluent households. Low/non-recyclers tend to be younger, low income and ethnic minority households. These low participation households are those described earlier as likely to suffer multiple constraints on recycling. 40% of households claim to be doing more recycling than in the past, though 9% say they are doing less. Those doing more are typically affluent, older households; those claiming to do less are most likely to be younger households living in flats.

Claimed high participation and increased recycling activity is most strongly associated with households saying they have kerbside collection. Nearly three-quarters of kerbside households claim to be high/medium recyclers and 59% claim increased recycling in the last few years. Even so, more than a quarter of kerbside households (28%) appear not to have been persuaded, and continue to do little or no recycling. Less than half of kerbside households claim to recycle as much as they can.

Claimed Recycling Habits – Materials

People appear to claim recycling participation in relation to a subconscious model of 'good behaviour'. The London surveys suggest that 'model' recycling behaviour is defined principally in relation to paper and glass, which are recycled regularly by around half of survey households. High recyclers tend to define themselves so principally on the basis of their habits for paper and glass.

Recycling participation is low for all other materials, and across all types of household (except clothing which is often disposed of outside the municipal waste stream). Only one in five London survey households regularly recycles anything other than paper and glass. (Self-reported) effectiveness of participation is variable, even amongst high participation households. Between 30 and 50% of regularly recycling households are not consistently recycling even the 'model' materials of paper and glass. Less than a quarter of committed recyclers recycle food and drink cans regularly, and regular plastics recycling is negligible, even amongst high recyclers. These findings emphasise the degree to which people's accepted norms about recycling will have to be transformed if recycling in London is to increase significantly.

The survey results suggest a need for both greater participation and effective recovery of materials for which there is an established 'norm' of behaviour, as well as a need to engage committed recyclers in recycling a wider range of materials.

Constraints on Recycling

While people *generally* think that more people in London should recycle, most survey participants have deep-seated beliefs why they *personally* could, or should not, do more. Myths about dumping of materials collected for recycling undermine some people's psychological commitment. A significant minority also believes that the local Council does not recycle all the materials that it asks for. Indeed, many people feel that their own recycling efforts are not matched by Councils 'doing their bit'.

For many people, environmental motivations to recycle are irrelevant. The reason that they do not recycle is that they simply do not think about it, or do not consider it of sufficient importance to justify the effort involved. Low and non-recyclers were most likely to hold these views. High recyclers find it easiest to fit recycling into everyday routines – but these are also the households most likely to have kerbside collection services, and they have feelings of moral responsibility about waste. Aspects of recycling that are felt to be 'too difficult' by a sizeable minority are storing recyclables and making special trips to bring banks. These constraints appear greatest for low and non-recyclers, who are most likely to live in flats and have to walk to bring sites.

Even self-declared committed recyclers, however, may be put off by the clutter and untidiness caused around the home by storing recyclables, as well as the time commitment involved in making special trips to bring sites. The way in which waste services are delivered communicates important messages, which affects individuals' commitment to participate in recycling. Knowledge about what to do is typically absorbed from what is visible in the local area, either through kerbside schemes or nearby bring banks. Many people do not think about recycling unless it is brought to their attention and very few actively seek out information on recycling services. Equally, the way in which recyclables are apparently managed sends important messages to ordinary people about the commitment of the local Council to recycling. Commonly expressed deterrents were overflowing bring banks and lack of evident separation on kerbside collections.

In general, the survey and group discussions suggest that Londoners' commitment to recycling is fragile. Very few people reject the recycling message outright and most people are aware, if only vaguely, that more needs to be done. However, recycling is a low priority for most households and is done most conscientiously where participation is made easy.

The research explored what ordinary people think can be done to reduce the amount of waste created, *via* a series of discussions about how people cook, and shop, and how they feel about packaging. Attitudes were also tested in the household survey. Product choice is dominated by perceptions of quality and cost – and is influenced by the fact that almost all households rely on supermarkets for their main food shop. As a result, special offers are influential and many people admit to buying more than they need because of offers.

People whose shopping habits are influenced by environmental concerns are in a minority, though a few take useful 'small steps' such as buying refills or choosing glass over plastic containers occasionally. Given households' generally low environmental motivation regarding shopping, it is not surprising that packaging is not an important issue for most people. The majority believes that packaging is impossible to avoid, principally because their eating, cooking and food shopping habits are largely fixed, and organised to deliver maximum convenience. Some high recyclers feel that consuming heavy packaging is acceptable so long as it can be recycled. Although people feel they cannot avoid packaging, many are nonetheless irritated by the amount of packaging of supermarket products. For some their irritation is related to environmental concerns; for others it is a belief that packaging deceives consumers about the products inside.

Re-use of products, including supermarket plastic bags, is minimal and not an issue in most households. Some parents donate (questionable quantities) of packaging to their child's school; most people claimed to pass on old clothes either to family or charity shops. In general, people in London do not feel that producing less waste is their responsibility, nor an issue over which they have any influence. Reduction of packaging is seen by most people as a job for retailers and manufacturers, who should be persuaded by regulation if necessary.

Options for Changing Behaviour

Although there is always likely to be a minority of households who will refuse to participate in recycling (perhaps as many as one in five), and similar numbers are resistant to being told what to do, the vast majority believe they can do more – with the right kind of help. 'Make recycling easy' was the overwhelming message from both the focus groups and household survey. For many people, this means making recycling as easy as throwing away normal refuse. Not surprisingly, therefore, provision of kerbside recyclables collections and more bring sites *close to home* were the two most popular interventions that people thought would motivate them to do more recycling – supported by 77% and 71% respectively.

Medium and low recyclers felt they needed to be reminded about recycling more directly. As well as specific information, visibility of bring banks acts as an important reminder for some. High recyclers want local Councils to take a wider range of materials. In reality, people tend to obtain most of their information on recycling by accident – from what they have seen around them. However, people *expect* information on recycling to be provided to them by the local Council, or environmental campaigning groups.

A common perception that recycling saves Councils money is one of several powerful myths about waste generally, and local government in particular, that influences how people respond to the idea of being asked to do more recycling. Long held views about the Council Tax make it difficult for people to think about how direct charging would differ from the present system, and how they might benefit. Charging is often perceived as imposing *extra* costs. It is seen as especially unacceptable by low income households and low recyclers, and unfair by a majority. People respond moderately positively to the idea of financial rewards for recycling. Both in relation to rewards and information, people were most comfortable with ideas that made sense in relation to established cultural norms – *e.g.* TV advertising, reward cards, web sites (for professionals).

People think that 'carrots' should come before 'sticks' – which would mean greater investment in making recycling easier (less effort) first. Some feel that greater, and demonstrable, commitment from local Councils should be demanded before people are punished for their own lack of commitment to recycling. If waste behaviour is to be transformed as required, building trust and dialogue with household customers of recycling services will be at least as important as providing extra facilities and new services.

Waste Service and Policy Implications

The London surveys provide important insights into the waste behaviour of households and identify ways in which people themselves think they can be motivated to change their habits. This was the primary purpose of the research. However, a number of implications for service interventions and for waste policy arise from the research. More than half of London households are most likely doing little or no recycling at present; only one in five households regularly recycles anything other than paper and glass; and minimisation is not on the agenda in most households. The scale of the challenge of meeting statutory recycling targets is thus clearly enormous. However, there were a number of positive signals from the survey. A majority – two-thirds of households – indicate that they could do more recycling, if provided with the right kind of help. Turning these good intentions into action will clearly be key to delivering more effective waste recovery. Before this can be done, however, policy makers and the waste industry will have to face up to some uncomfortable truths about people's attitudes and behaviour.

The surveys show that there are, as suspected, real differences between types of household in their commitment to recycling, and that these differences are associated with age and social class factors. A particularly difficult truth to face up to is that there is a resolute minority that refuses to recycle – amongst all social classes and age groups, but represented more strongly amongst 'socially excluded' households. To maximise the chances of engagement from the public, messages and interventions will need to be targeted differentially to appeal to the different values and motivations of different groups (an approach which may clearly be politically sensitive).

However, another important point to acknowledge is that, because of the way services are organised, recycling is made easiest for those who already have the greatest inclination to recycle – broadly middle class and older households – and hardest for those with the weakest motivation. Also unavoidable is the fact that many people, from all social backgrounds, are deeply distrustful of local Councils, and that people's experiences of other local authority services and the Council Tax are important barriers to them being able to think rationally about waste issues. The most important barrier, however, is undoubtedly the widespread perception that recycling (and minimisation) is difficult – meaning *difficult to fit into my life* rather than *difficult to use in practice*.

Making recycling more convenient will therefore be a primary requirement if recycling targets are to be met – including more kerbside provision and *near home* bring banks. Involving the customers (households) in debate about what provision is right for them will also be an important component of delivering services people want to use. However, delivering a larger number of services to more people will not be enough alone to secure as much action on recycling and minimisation as is required. Neither will simply telling people that services exist. People will need to be persuaded that recycling and minimisation are 'normal' things to do, not just a 'good' thing. They will also need to be persuaded that everyone is making an effort, so that participation is fair, and that local authorities are prepared to 'do their bit'. This means not only providing quality services but also being seen to do it – either by providing feedback or indirectly by the way in which services are delivered. Drawing on the experience of other situations where individuals and companies are being persuaded to adopt more sustainable behaviour suggests that building gradually on small 'successes', rather than attempting to engineer an immediate and giant leap, are more persuasive tactics.

Policy makers at local and London levels are well placed to smooth the required transformation in behaviour, including: co-ordinating and advertising the message; representing Londoners' views on packaging to producers; partnering the waste industry to overcome barriers on the ground; and generally to provide an arena for the exchange of best practice and experience. As regards financial incentives, our interpretation of the surveys is that London households on the whole are not yet ready for charging or direct financial incentives. This is not to say that those reward schemes that have been introduced in London and elsewhere are not useful, nor that they may not be influential once the majority has engaged with the idea of less waste, but simply that they may not be an immediate solution to transforming behaviour.

9   Conclusions

The reasons for wishing to manage waste well have changed in recent years, and what constitutes effective waste management has also changed. We are now in an age where the aspects of public and occupational health and safety have been joined by policy needs on environmental protection, resource consumption and sustainability.

The challenge to de-couple economic growth from waste creation has not yet been met and there are increasing policy instruments in place which are squeezing the gap between what happens now and what needs to happen in future. Waste management practices are not too sophisticated to be addressed using current technologies, providing policy-makers can define appropriate goals. The main objective in the medium term will be to decide who must pay for the new world of integrated waste management; how to strike a balance between charging us all as tax-payers or as consumers.

Many of the mechanisms for progress are now being put in place, and we can clearly see that policy instruments are pushing things in the right direction. Technical solutions exist and new policies should enable these to flourish.

However we as society decide to pay the bills incurred by our activities which generate waste, it is apparent that the heart of the challenge will be to ensure that as many people as possible change their behaviour, to reduce the impacts of our production and consumption upon the environment.

Health Risks of Materials Recycling Facilities

TONI GLADDING

1 Introduction

There are several types of Materials Recycling Facilities (MRFs, also known as Materials Recovery Facilities, or Materials Recycling Factories) currently in operation both in the UK and in Europe. These can generally be divided into those that sort and process construction and demolition waste, and those used to sort and process source-segregated household and commercial waste. This review will mainly concentrate on the latter, and most popular, type of MRF. MRFs that deal with household and commercial waste are defined as:

'A central operation where source-segregated, dry, recyclable materials are sorted, mechanically or manually, to market specifications for processing into secondary materials'[1]

A full discussion of the use, waste collection systems, design and operation, *etc.* surrounding the MRF is given by the Institute of Wastes Management.[1]

Governmental waste strategies, based on a variety of drivers from Europe including the Landfill Directive, have increased demand for recycling of packaging-related materials (plastics and glass) and paper. The Landfill Directive (1999/31/EC) places stricter controls on landfill practices, and requires member states to divert from landfill substantial quantities of solid wastes, a reduction to only 35% of the amount disposed of through landfill in 1995 by 2020. This is an onerous target, and given that waste arisings are increasing by approximately 3% per annum recycling is and will become ever more important in meeting these targets. In addition, the Waste Strategy released by the Government in 2000 set targets for recycling, and estimated that between 100 and 300 new MRFs with an average capacity of 40 000 tonnes p.a. would be required as a result. To a lesser extent, the continued increases in landfill taxes and the establishment of the Waste Resources Action Programme will also further increase tonnages entering MRFs.

---

[1] Institute of Wastes Management, *Materials Recovery Facilities*, IWM Business Services Ltd., 2000, ISBN 0 902944 57 6.

Issues in Environmental Science and Technology, No. 18
Environmental and Health Impact of Solid Waste Management Activities
© The Royal Society of Chemistry, 2002

As a result of these changes in targets, the number of MRFs are increasing rapidly in the UK,[1] from only half a dozen in the late eighties and early nineties, to somewhere in the region of 100 facilities at present. The average number of employees in MRFs is between 11[1] and 19[2] workers (range 5–49,[1] to range 4–40[2]). Therefore, there are in the region of 1100–1900 workers handsorting waste in England and Wales at present, and this number is set to rise to 400 MRFs employing some 4400–7600 workers full-time.

In view of the expanding work force employed in MRFs, studies investigating potential health effects on MRF workers are important. This chapter is divided into two parts. In the first part, it reviews previous work that has been carried out on MRF workers both in the UK and abroad. Secondly, it outlines a recent study to assess the relationship between various exposures and work-related symptoms and effects to MRF workers funded by a European BIOMED2 programme and the Environment Agency for England and Wales.[2] This research included exposure measurements at eleven MRFs in England and Wales concentrating on air quality, noise and electromagnetic frequencies. At nine MRFs measurements of the health of workers *via* self-reported questionnaires of symptoms and more objective measurements such as blood counts and lung function testing were carried out. This research compared these eleven different MRFs in terms of technology, size, materials accepted, method of collection of materials, residue rates and situation for evaluation. This research is the most comprehensive programme of data collection in MRFs to date in terms of exposures within MRFs.

What Are the Issues in MRFs?

Household waste contains a diversity of materials and therefore potentially numerous hazards, with sheer volume exacerbating any difficulties. Previously, waste management was progressively more mechanised, and these hazards have been contained from householder storage to collection through 'traditional' disposal routes (landfill and mass burn incinerators). MRFs effectively reverse this distancing of workers from the waste materials, and bring waste into closer contact with operatives due to handsorting of materials. These hazards can be grouped into three main areas, shown in Table 1.

Manual handling of materials and the ergonomic aspects of materials handling handsorting are the main physical hazards, followed by the potential for accidents, *e.g.* cuts (broken bottles), broken limbs *etc.* especially during interaction with heavy machinery and movement of vehicles. Many MRFs are also vulnerable to potential fires. Noise and vibration are present in MRFs due to the use of various sorting and baling machinery. Electromagnetic fields (EMFs) are potentially important due to the use of ferrous and non-ferrous separation equipment. Although there are no available figures in the UK for the magnitude of these problems, reports from the USA suggest that these are not insignificant issues within MRFs. In 1993 when MRFs were first introduced in the USA a

[2] T.L. Gladding, *An Assessment of the Risks to Human Health of Materials Recovery Facilities: A Framework for Decision Makers*, Final Report Environment Agency Contract No: P1-214, 2002.

Table 1 Hazards in MRFs

Physical	Chemical	Biological
Manual handling	Hazardous waste residues	Airborne microorganisms
Ergonomics	Hazardous waste vapours/aerosols	Contaminated sharps
Accident, transport, fire	Heavy metals, *e.g.* lead, mercury *etc.*	Contaminated sharp edges
Noise and vibration Electromagnetic frequencies	Volatile Organic Compounds	Total and respirable dust

'health and safety manual' was produced encompassing such areas as manual handling, traffic, noise and fire.[3]

Chemical hazards include vapours and residues from household hazardous waste (HHW), *e.g.* garden chemicals, wood preservatives, paints, cleaning materials *etc.* Heavy metals are included in this category due to the possibility of exposure to cadmium and mercury from batteries in HHW, first investigated in MRFs in Denmark, showing some presence of mercury and lead.[4] Volatile organic compounds (VOCs) are produced when waste is degrading, *e.g.* organic sulfur compounds are thought to contribute to complaints of nausea, irritation and intestinal problems experienced by some operatives.[5]

Biological hazards have caused most concern in MRFs. Collection and separation of household waste generates organic dusts. These include airborne bacteria and fungi (bioaerosols) and their cell wall components. Microbial cell wall components are an important constituent in organic dusts.[6] Among the best known of these are bacterial endotoxins (a cell wall component in Gram-negative bacteria). Relationships between the amount of endotoxin in different environments and respiratory symptoms, spirometry changes and increased inflammatory markers have been reported.[7] In addition, $(1\rightarrow3)$-β-D-glucans, present in the cell wall of fungi also have a number of toxic properties.[8]

Dusts generated in waste facilities could also include airborne viruses.[9] Viable or live microorganisms are implicated in infection and allergy, and pathogenic species such as *Aspergillus fumigatus* are of some concern in composting.[10] Viable

[3] L. Fredrickson, *Safety in Recycling Facilities – A Resource for Operators*, Minnesota Pollution Control Agency, 1992.

[4] T. Sigsgaard, J. C. Hansen, P. Malmros and J. V. Christiansen, *Work Related Symptoms and Metal Concentration in Danish Resource Recovery Workers*, Biological Waste Treatment, James & James, London, 1996.

[5] C. K. Wilkins, Gaseous Organic Emissions from Various Types of Household Waste, *Ann. Agric. Environ. Med.*, 1997, **4**, 87–89.

[6] R. Rylander and R. R. Jacobs, *Organic Dusts: Exposure, Effects and Prevention*, CRC Press Inc., Boca Raton, 2000 Corporate Blvd., Fl, USA, 1994, ISBN 0-87371-699-X.

[7] R. Rylander, Evaluation of the Risks of Endotoxin Exposures, *Int. J. Occup. Environ. Health*, 1997, Supplement to **3** (1), S32–S36.

[8] R. Rylander, Indoor Air-Related Effects and Airborne $(1\rightarrow3)$-β-D-Glucan, *Environ. Health Perspect.*, 1999, **107** (S3), 501–503.

[9] A. Pforrmann and G. van den Bossche, Occurrence and isolation of human enteroviruses from the air of waste removal and disposal plants, *Zentralbl. Hyg. Umweltmed.*, 1994, **196**(1), 38–51.

[10] M. N. Kramer, V. P. Kurup and J. N. Fink, Allergic Bronchopulmonary Aspergillosis from a Contaminated Dump Site, *Am. Rev. Respir. Dis.*, 1989, **140**, 1086–1088.

microorganisms are measured in colony forming units (cfu); their viability is measured by their growth in a laboratory. However, total numbers of microorganisms, including those alive and dead, are causing particular concern in waste management.[11] When microorganisms are aerosolised, their viability decreases. However, total number of cells can still potentially cause airways irritation.[11] Endotoxin and $(1\rightarrow3)$-β-D-glucan exist whether the bioaerosol is alive or dead.[7,8] They are implicated in fever, flu-like symptoms, headaches, excessive tiredness and joint pains (termed 'Organic Dust Toxic Syndrome') and gastrointestinal problems.[7,8] These symptoms have been reported in studies on waste sorting facilities.[11,12]

There are currently no occupational exposure limits for microorganisms, endotoxin or glucan. Very little research has been carried out on what constitutes 'safe' levels of bioaerosols to which an individual can be exposed with respect to facilities that deal with waste. This general lack of data also means it is very difficult to draw on past studies and provide definitive conclusions. Unfortunately, satisfactory dose–response data are not available. Therefore, related legislation and guidance, and research from a variety of workers are used as a guideline for exposure, as seen in Table 2.

There are no recognised exposure limits for bioaerosols, and as stated earlier no dose–response data. However, the Danish Working Environment Service propose that levels in excess of 1×10^6–10^9 cfu m^{-3} could cause respiratory problems.[13] Other authors state a more conservative 10^4 cfu m^{-3} as a guidance level for waste management facilities based on work in waste facilities, but this is not supported by dose–response data (and it is recommended as a guideline only).[14] In terms of viable microorganisms, it has been reported concentrations in excess of 10^6 cfu m^{-3} have been found to lead to hypersensitivity pneumonitis (allergic alveolitis) complaints, *e.g.* Farmers' Lung.[16] Other sources state that natural concentrations of microorganisms routinely range from 1000 to 100 000 cfu m^{-3} air.[15,17] However, without dose–response data, concentrations of either viable or non-viable microorganisms, endotoxin or glucan are difficult to evaluate in terms of whether MRFs will experience problems.

A review showing concentrations in MRFs is outlined below. Due to the ongoing debate on bioaerosol exposure in waste handling facilities, many

[11] O. M. Poulsen, N. O. Breum, N. Ebbehøj, A. M. Hansen, U. I. Ivens, D. V. Lelieveld, P. Malmros, L. Matthiasen, B. H. Nielsen, E. M. Nielsen, B. Schibye, T. Skov, E. I. Stenbaek and C. K. Wilkins, Sorting and Recycling of Domestic Waste. Review of Occupational Health Problems and their Possible Causes, *Sci. Total Environ.*, 1995, **168**, 33–56.

[12] T. Sigsgaard, P. Malmros, L. Nersting and C. Petersen, Respiratory Disorders and Atopy in Danish Refuse Workers, *Am. J. Respir. Crit. Care Med.*, 1994, **149**, 1407–1412.

[13] N. O. Breum, H. Wurtz, U. Midtgaard and N. Ebbehøj, Dustiness and Bioaerosol Exposure in Sorting Recyclable Paper, *Waste Manag. Res.*, 1999, **17**, 100–108.

[14] J. Lavoie and R. Alie, Determining the Characteristics to be Considered from a Worker Health and Safety Standpoint in Household Waste Sorting and Composting Plants, *Ann. Agric. Environ. Med.*, 1997, **4**, 123–128.

[15] C. Y. Rao, H. A. Burge and J. C. Chang, Review of Quantitative Standards and Guidelines for Fungi in Indoor Air, *J. Air Waste Manag. Assoc.*, 1996, **46**, 899–908.

[16] J. Lacey and J. Dutkiewicz, Bioaerosols and Occupational Lung Disease, *J. Aerosol. Sci.*, 1994, **25**, 1371–1404.

[17] C. S. Cox and C. M. Wathes (ed.), *The Bioaerosols Handbook*, Lewis Publications Ltd., 1995.

Table 2 Recommended exposure limits for viable microorganisms

Known sources of emissions	Agency	Description	Recommended limit	Sizes	Potential health effects	Notes
Natural, industrial and farming, putrescible biodegradation, compost	Lavoie et al. (1991) (USA)[14] Breum et al. (1999)[13] (DK) Rao et al. (1996)[15] (various standards reviewed)	No legal standards – only for guidance From reviews, consensus and surveys in non-contaminated indoor environments	10^4 cfu (colony forming units) m^{-3} 100–1000 cfu m^{-3} (non-contaminated indoor environment)	All: coarse (10+ μm) to ultra-fines (<2.5 μm)	10^{6-8} cfu m^{-3} and above known to cause allergic alveolitis. Other health effects depend on particular micro-organisms present	Research based effects and limits at present, absence of standard protocols, little data on human dose–response relationship

European countries have initiated regulations for their workers. Germany has introduced health regulations for all of its DSD plants and 11 000 workers.[18] The Netherlands has discussed setting an occupational exposure limit for endotoxin at 4.5 ng m^{-3} as an 8-hour time weighted average.[19] Research has indicated that levels of endotoxin (and glucan) in excess of 10 ng m^{-3} are thought to cause airways inflammation in individuals with previous asthma or eczema type disorders.[7,8]

Finally, many MRFs have encountered contaminated sharps in the UK, which are usually from domestic sources, *e.g.* diabetic users. Contaminated sharp edges refer to glass or metals that may lead to infection or disease. Of particular interest are tetanus, hepatitis (various strains) or less likely HIV.

Many countries are now recognising potential hazards in MRFs, particularly Denmark, The Netherlands and Germany, but also to a lesser extent Sweden, Norway, Finland, the USA and Canada. The USA discussed potential risks when MRFs were first introduced, and carried out some assessments.[3,20] An early United States Environmental Protection Agency publication recommended that various occupational aspects in MRFs were investigated in more detail, particularly exposures to bioaerosols and dusts.[20] However, little research followed these publications highlighting the areas of potential risk. Most of the research concerning handsorting has been carried out in Europe, as outlined below.

2 Previous Research Concerning Waste Handsorting

Waste management has previously been an under-researched occupation. In the UK, accident and illness have traditionally not been separated in national health and safety statistics. Many studies are concentrated on mixed waste sorting facilities, many of which produce Refuse-Derived Fuel (RDF) for incineration. Mixed waste sorting facilities are distinct from MRFs in that the quality of recyclates recovered is lower than in a positively managed waste stream, as outlined in a recent publication on MRFs.[1] At present there are only one or two facilities based around this concept in the UK. Results from studies undertaken at mixed waste sorting plants are thought to show higher concentrations of dust, bioaerosols and metals than those taken at source-segregated facilities. Table 3 illustrates studies on mixed waste facilities.

Early studies identified exposures at mixed waste and RDF plants that could be of concern.[21–24]

[18] TBRA, *Waste Sorting Plants: Guidance for Protection Measures*, Ministry of Work, Bundesarbeitsblatt, Germany, TBRA 210, 1999.

[19] Dutch Expert Committee on Occupational Standards, *Health-based Recommended Occupational Exposure Limit for Endotoxins*, Gezondheidsraad Postbus 1236, 2280 CE Rijswijk, The Netherlands, 1997.

[20] United States Environmental Protection Agency, *Public Health, Occupational Safety and Environmental Concerns in Recycling Operations*, EPA/600/R-93/122, 1993.

[21] L. F. Diaz, L. Riley, G. Savage and G. J. Trezek, Health Aspect Considerations Associated with Resource Recovery, *Compost Sci.*, 1976, **17** (3), 18–24.

[22] P. J. Constable and D. J. Ray, Consideration of Health Hazards Associated with the Recycling of Household Waste, *Environ. Health*, 1979, **87** (9), 193–195.

[23] D. Mozzon, D. A. Brown and J. W. Smith, Occupational Exposure to Airborne Dust, Respirable

Table 3 Sorting of unseparated waste

Author	Country	Facility-type	Contaminants and Concentrations
Diaz et al. (1976)[21]	USA	Resource recovery facility	Aluminium/cadmium and asbestos were undetectable, iron 374.1 μg g^{-1} and lead 235 μg g^{-1} (in particles sampled). Bacteria not over 0.36×10^3 cfu/ft air. In the waste faecal coliforms reached 10^6, faecal streptococci 10^6 cfu m^{-3}.
Constable et al. (1979)[22]	UK	Mixed waste sorting/RDF plant	*Penicillium* to 5.8×10^4 cfu m^{-3}, and bacteria high (not specified). Few *Aspergillus*, *Cladosporium*, Actinomycetes.
Mozzon et al. (1987)[23]	USA	RDF plant	Lead up to 2.1 mg m^{-3} at the precipitator cleaner, though more often at 0.003 mg m^{-3}. Cadmium peaking at 0.32 mg m^{-3} in the same area, and in other parts 0.003 mg m^{-3}. Asbestos and PCBs were not found in any samples.
Crook et al. (1987)[24]	UK	RDF plant	Bacteria up to 10^6, fungi 10^7 and actinomycetes 10^5 cfu m^{-3}. Gram-negatives and pathogenic microorganisms caused concern.
Malmros (1988)[25] Sigsgaard (1990)[26] Malmros (1992)[27] Malmros et al. (1994)[28]	Denmark	Mixed waste sorting to produce RDF pellets	Viable counts showed microorganisms to 20000 cfu m^{-3} in reception and during manual sorting. Endotoxins were highest at the RDF press, 0.99 μg m^{-3}. Rebuilding of the plant reduced these to 8400 cfu m^{-3} as the highest reading at the magnet and 0.11 μg m^{-3} during sorting respectively.
Rahkonen (1992)[29]	Finland	Mixed waste sorting	Lead to 0.26 μg m^{-3} and cadmium to 0.09 μg m^{-3} at the sorting belt. Bacteria to 1.4×10^4, fungi to 2.5×10^4 cfu m^{-3}. Total dust to 38 mg m^{-3}, endotoxin to 30 ng m^{-3}.
Pfirrmann et al. (1994)[30]	Germany	Waste removal plant	Viral infectivity in 12 of 36 samples, belonging to the family Picornaviridae.
Marchand et al. (1995)[31]	Canada	Mixed waste sorting	Bacteria to 5.2×10^5, Gram-negatives to 7.9×10^3, fungi to 7.2×10^3 cfu m^{-3}.
Jager et al. (1995)[32]	Germany	'Garbage' sorting	Bacteria to 1.4×10^4, Gram-negatives to 7.2×10^3, fungi to 8.4×10^4 cfu m^{-3}.
Streib et al. (1996)[33]	Germany	Mixed waste sorting	Airborne natural and artificial fibres found occasionally. Cadmium, mercury and nickel in the range of their natural concentrations in urban areas, lead found in excess. Microorganisms 6.9×10^5 cfu m^{-3}, fungi 6.6×10^4 cfu m^{-3}, with 90% below 7 μm. Exposure limit for dust of 6 mg m^{-3} exceeded for short periods.
Van Tongeren et al. (1997)[34]	N.L.	Resource recovery facility	Dusts up to 14.3 mg m^{-3} during manual separation of waste, endotoxin from 32 ng m^{-3} to 131.1 ng m^{-3}. Microorganisms to 10^6 cfu m^{-3} during tipping.
Mahar et al. (1999)[35]	USA	Two RDF plants	Total dust geometric mean 0.50 mg m^{-3}, endotoxin 2.9 ng m^{-3}, 6.8×10^5 cells m^{-3} total microorganisms.

In terms of the health of workers in mixed waste sorting facilities, the most reported case was of a plant in Denmark. In Denmark manual and mechanical waste sorting plants have been in operation since autumn 1986. One of the first, the 4S plant was a mechanical sorting plant receiving up to 10 000 t p.a. of mixed household and industrial waste, employing 20 people, and was equipped with a manual sorting line. Within months of opening five operatives with respiratory ailments were recorded, and a study of working environments within these plants was initiated by the Danish National Environmental Protection Agency.

Of 15 exposed operatives, five were asthmatic, but others were also exhibiting flu-like symptoms (possibly allergic alveolitis) eye and skin irritation, fatigue and occasional nausea.[25] Microbial decomposition activity and endotoxins were suspected as a cause of these effects. Inspections at the site indicated proliferation of dust and food waste among 'sortable' materials; accumulated wet refuse was sometimes mixed with material for re-use. Rebuilding of the plant was therefore undertaken, including encapsulation of conveyor belts and a central vacuum cleaning device. Further initiatives included new ventilation systems and microbial control programmes. Compressed air used to clean the facility at the end of the day, thought to contribute to the aerosolisation of microorganisms, was stopped.[25]

Medical studies of operatives from the 4S plant showed that eight operatives became ill within seven months of starting. In total, nine cases of occupational disease among the original fifteen exposed operatives were reported.[26] Eight similar cases occurred between August 1986 and March 1987; the ninth case occurred in September 1988. The first two operatives to become ill were involved in cleaning and handsorting. All symptoms began with eye irritation and sore throat, followed by respiratory symptoms, including chest tightness and exercise-induced dyspnoea. The ninth case arose when safety regulations were disregarded.[27] Five operatives were reported as showing abnormal peak flow recordings of more than 20% variability after work, two with a history of fever and flu-like symptoms, and four with positive titres of precipitins against crude dust from the plant.[26] This was subsequently updated to three suspected cases of Organic Dust Toxic Syndrome.[28] Eight of the nine cases were subsequently diagnosed with bronchial asthma, verified by a variation in peak flow of more than $100 \, l \, min^{-1}$.[28] It was also reported that a single aetiological agent was not found, radio-allergo sorbent skin test (RAST) analyses were negative and immunoglobulin E (IgE) were normal. They also ruled out infection, as leucocyte

Quartz and Metals Arising from Refuse Handling, Burning and Landfilling, *Am. Ind. Hyg. Assoc. J.*, 1987, **48** (2), 111–116.

[24] B. Crook, S. Higgins and J. Lacey, *Airborne Micro-organisms Associated with Domestic Waste Disposal*, Final Report to the HSE, Contract Number: 1/MS/126/643/82, 1987.

[25] P. Malmros, *The Working Conditions at Danish Sorting Plants*, ISWA Proceedings, Academic Press Ltd., 1988, Vol. 1, pp. 487–494.

[26] T. Sigsgaard, Respiratory Impairment among Workers in a Garbage-Handling Plant, *Am. J. Ind. Med.*, 1990, **17**, 92–93.

[27] P. Malmros, *Get Wise on Waste – A Book about Health and Waste-Handling*, Danish Working Environment Service, 1992.

[28] P. Malmros and P. Jonsson, Wastes Management: Planning for Recycling and Workers' Safety, *J. Waste Manag. Resource Recovery*, 1994, **1** (3), 107–112.

counts (number of white blood cells) were normal.[28]

In spring 1989 six had dyspnoea on exertion and three had positive histamine-provocation tests; seven had left the plant. The nine cases were registered with the National Board of Industrial Injury from the original plant, and seven were accepted as having occupational diseases.[27] It was finally reported that eight had bronchial asthma, one chronic bronchitis and one allergic alveolitis. Eventually, seven changed jobs, but only two out of the seven were free of symptoms after two years away from the plant.[28]

The case in Denmark remains the most investigated in terms of health of workers sorting mixed waste. Later studies identified that exposures were of concern in such facilities, but did not report the health effects seen in Denmark.[29-35]

There are fewer studies on the type of MRF now being developed on a large-scale in the UK – those that separate segregated materials. In 1995 six MRFs throughout the USA (New York, Maryland, New Mexico, Connecticut, Minnesota and Florida) were researched under a USEPA programme to consider environmental, economic and energy impacts. This was an Interagency Energy and Environmental Research Report (IEERR) including representatives from USEPA and the US Department of Energy.[36] The results of this study, and others that have also examined MRFs are outlined in Table 4.

Among the first studies concerning MRFs concentrations of bioaerosols were reported at 10^5 cfu m^{-3} or lower and relatively low levels of endotoxin and dust at six MRFs.[37] These results indicated that if waste was of good quality and if reception halls for unsorted waste are separated from operatives, concentrations of airborne microorganisms would be contained. Bioaerosols were reported to increase as quality of waste deteriorated and activity increased, which in turn may have an impact on operative's health, although this was not measured.

A later study in Denmark compared cross-sectional studies of 750 operatives in textile mills, recycling plants (including paper sorting), and a wet paper producing

[29] P. Rahkonen, Airborne Contaminants at Waste Treatment Plants, *Waste Manag. Res.*, 1992, **10**, 411–421.

[30] A. Pfirrmann and G. vanden Bossche, Occurrence and Isolation of Human Enteroviruses from the air of Waste Removal and Disposal Plants, *Zentralbl. Hyg. Umweltmed.*, 1994, **196** (1), 38–51.

[31] G. Marchand, J. Lavoie and L. Lazure, Evaluation of Bioaerosols in a Municipal Solid Waste Recycling and Composting Plant, *J. Air Waste Manag. Assoc.*, 1995, **45**, 778–781.

[32] E. Jager, H. Ruden and B. Zeschmar-Lahl, Air Microbial Burden at Garbage Sorting Facilities, *Zentralbl. Hyg. Umweltmed.*, 1995, **197**, 398–407.

[33] R. Streib, K. Botzenhart, K. Drysch and A. W. Rettenmeier, Dust and Microorganism Count at Delivery, Sorting and Composting of Home Refuse and Home Refuse-like Industrial Waste, *Zentralbl. Hyg. Umweltmed.*, 1996, **198**, 531–551.

[34] M. Van Tongeren, L. Van Amelsvoort and D. Heederik, Exposure to Organic Dusts, Endotoxins, and Microorganisms in the Municipal Waste Industry, *Int. J. Occup. Environ. Health*, 1997, **3**, 30–36.

[35] S. Mahar, S. J. Reynolds and P. S. Thorne, Worker Exposure to Particulates, Endotoxins and Bioaerosols at two Refuse-Derived Fuel Plants, *Am. Ind. Hyg. Assoc. J.*, 1999, **60**, 679–683.

[36] Interagency Energy and Environmental Research Report (IEERR), *Environmental, Economic, and Energy Impacts of Materials Recovery Facilities – A MITE Programme Evaluation*, EPA/600/R-95/125, NREL/TP430-8130, 1995.

[37] L. Nersting, P. Malmros, T. Sigsgaard and C. Petersen, Biological Health Risk Associated with Resource Recovery, Sorting of Recycle Waste and Composting, *Grana*, 1991, **30**, 454–457.

Table 4 Sorting of separated waste

Author	Country	Facility-type	Contaminants and Concentrations
Nersting et al. (1991)[37]	Denmark	Source segregated waste	Bioaerosols from 6×10^2 to 4.7×10^4 cfu m^{-3}. One sorting hall with no ventilation/temporary shielding had fungi levels of 1.4×10^5 cfu m^{-3} and 5.4×10^5 cfu m^{-3} Gram-negative bacteria were found when large quantities of waste were sorted. Endotoxins up to 14.41 ng m^{-3}. Dust concentrations less than 5 mg m^{-3}.
Sigsgaard (1993)[38]	Denmark	Source segregated waste compared to paper and composting operatives	Fungi levels around 1.4×10^4 cfu m^{-3} and bacteria between 5×10^3–10^5 cfu m^{-3} in waste handling plants. Significantly higher endotoxin concentration was found in waste handling plants compared to paper sorting plants.
IEERR (1995)[36]	USA	Six MRFs	Environmental measurements Pb 0.07 μg m^{-3}, Hg 0.006 μg m^{-3}. Occupational measurement silica 0.12 mg m^{-3} and metals all below 0.01 μg m^{-3}. Total dust up to 2.50 mg m^{-3}, respirable 0.57 mg m^{-3}. Peaks of microorganisms at different sites seen in Table 5. Suspended particulates up to 122.75 μkg m^{-3}.
Sigsgaard et al. (1996)[4]	Denmark	Source segregated waste compared to paper and composting	Lead up to 3.9 μg l^{-1}, mercury 2.3 μg l^{-1} in the blood. Cadmium up to 3.6 μg l^{-1} in blood of waste workers compared with 1.7 μg l^{-1} in controls. Total dust highest in waste handling plants, at 0.74 mg m^{-3}. Waste handling and compost plants showed the highest viable counts of bacteria and fungi up to 83×10^4 cfu m^{-3}.
Gladding et al. (1997)[39]	UK	Source segregated MRFs	Bacteria and fungi to 2.5×10^5 cfu m^{-3} with total dust levels to 18 mg m^{-3}.
Kivranta et al. (1999)[40]	Finland	Source segregated sorting	Viable fungi, bacteria and Gram-negative bacteria to 10^5 cfu m^{-3}, VOCs peaking at 3000 μg m^{-3} considered to be the limit for discomfort.
Lavoie et al. (2001)[41]	Canada	Segregated materials	Bacteria to 2.1×10^4 cfu m^{-3}, Gram-negative bacteria to 3.2×10^3 cfu m^{-3}, fungi to 1.4×10^4 cfu m^{-3}. CO_2, CO, NO and NO_2 not measured in significant amounts. EMFs low, noise exceeded 90 dB(A) in one plant. Ergonomics a possible risk factor for MRF workers.

plant and a water supply plant.[38] This included interviews, lung function testing, peak flow monitoring, skin prick tests and serological tests. Viable airborne bacteria (including Gram-negative bacteria) and fungi were collected. In these industries a negative association was found between different markers of atopy and increasing levels of endotoxin, which indicated a healthy worker selection. This was further emphasised by operatives in this study who left recycling after a short period of employment because of asthma symptoms.

The most common symptoms in waste handling were itching eyes (26.7%) and sore or itching throat (20.5%). Toxic alveolitis was found in excess among waste handling (14%), composting and paper production (23%) operatives, but allergic alveolitis was not observed in any of these. Non-specific chest tightness was more common in waste handling. Lung function testing revealed significantly lower FVC% in paper sorters compared to controls and waste handling operatives, and a significantly higher $FEV_1/FEC\%$ among paper sorters compared to waste handling operatives, which was linked to dust rather than endotoxin or fungi levels. Peak flow monitoring showed greater variability in waste handling operatives and in paper production plants. Chronic effects on lung function parameters were not detected, but may be partly explained by the 'newness' of recycling. Serological testing revealed significantly lower IgE serum concentrations among paper sorting and waste handling operatives than controls (water supply operatives). Irritative skin symptoms were significantly more common among waste handling operatives than controls, but skin prick testing revealed no association between work-related asthma in recycling. There were no allergens in the environment to which operatives reacted.[4,38]

When statistical ORs were used to construct a multiple logistic regression analysis adjusting for smoking, age and atopic predisposition it revealed the OR for chest tightness and ODTS to be significantly increased among waste handling operatives [OR95% *i.e.* 5.43 (2.01–14.6)]. Atopic disposition also had an increased OR for 'Organic Dust Toxic Syndrome' [2.28 (1.07–4.84)]. Additionally, chronic bronchitis and chronic dry cough were associated with smoking OR [OR 4.53 (1.56–13.77) and 2.28 (1.39–3.7) respectively]. Waste handling operatives had significantly increased rates of work-related chest tightness, influenza feeling or fever, and mucus membrane irritative symptoms. Also, gastrointestinal symptoms such as nausea, work-related vomiting and diarrhoea were more often reported among waste handling operatives than among controls. It was suggested this was related to enterotoxins produced by microorganisms in the environment.[12,38]

The largest study previously in the United States involved six MRFs.[36] These six MRFs covered a range of manual and mechanical segregated waste sorting techniques. Measurements were made occupationally for dust, bioaerosols, heavy metals and noise. Metals results showed that for occupational measurements metals such as silica were measured at 0.12 mg m^{-3} (more often below 0.01 mg m^{-3}) well below state guidelines. Metals such as arsenic were measured up to < 0.0002 mg m^{-3}, aluminium 0.0115 mg m^{-3}, chromium < 0.0018 mg m^{-3}, lead < 0.0018 mg m^{-3}, nickel < 0.0018 mg m^{-3}. As these concentrations were considered

[38] T. Sigsgaard, *Organic Dust and Respiratory Symptoms in Selected Industrial Environments*, Institut for Epidemiologi og Socialmedicin, Rapport Nr. 5, 1993.

T. Gladding

Table 5 Viable
Microorganisms
($\times 10^4$ cfu m^{-3}) in
MRFs[35]

Site	Gauze Fungi/Bact.	Fungi	Bacteria RT_1	Bacteria 56°C
Ambient u/w*	N/A	0.7	0.6	0.03
Ambient d/w*	N/A	0.7	0.3	0.4
Tipping floor	2.0/15.0 (100)	0.9	0.9	0.04
Pre-sort	6.9/9.0	0.9	0.9	0.01
Plastic lines	19.0/170.0 (6300)	0.9	0.9	0.03
Glass lines	0.1/0.5	0.9	0.9	0.07
Paper lines	0.3/0.08	0.5	0.2	0.0094
Baler	160/92.0	0.5	0.6	0.02
Bale storage	N/A	0.06	0.06	0.0012
Lunch room	0.1/0.4 (8000)	0.1 (0.9)	0.09 (0.1)	0.0085

* where ambient stations not specified, treated as downwind. RT_1: room temperature.
(Figures in brackets from MRF in New York – markedly different from all other five)

very low measurements were discontinued. The highest concentration of mercury measured at the six MRFs was 0.005 mg m^{-3} (sorting tin and aluminium) in personal measurements. However, in common with previous measurements of heavy metals, generally results were considered very low or undetectable. It was concluded that measurements of heavy metals in an occupational context could need further evaluation.

The same study undertook various community measurements for heavy metals: lead up to 0.08 upwind and 0.07 μg m^{-3} downwind, mercury from undetectable to 0.006 mg m^{-3}. For other chemicals, such as PCBs and pesticides, measurements were barely within detectable limits at one MRF; therefore further measurements were discontinued. They concluded that there was no significant impact by a MRF on the surrounding community from these parameters.

Total suspended particulates were also measured at the six MRFs, both occupationally and environmentally. The environmental concentrations measured ranged from up to 122.75 μg m^{-3} upwind (one measurement of 301.06) and 138.90 μg m^{-3} downwind. PM$_{10}$ reached 107.06 upwind and 335.40 μg m^{-3} downwind. For occupational measurements, personal total dust recorded levels up to 2.50 mg m^{-3}, but was more often below 1 mg m^{-3}, respirable dust was measured up to 0.57 mg m^{-3}, but was more often below 0.2 mg m^{-3}.

A range of measurements were also taken for bioaerosols. In addition to viable air sampling, gauze wipes of operatives' hands were taken (although the air sampler was unnamed, an Andersen Sampler seems likely from descriptions in the text). Results of airborne bioaerosols are shown in detail in Table 5.

Environmentally, concentrations did not differ significantly between upwind and downwind measurements showing little impact on surrounding communities. Speciation of microorganisms was carried out using gas chromatography and selective media. Bacteria detected most commonly were *Bacillus, Curtobacterium, Arthrobacter, Streptomycetes* and some *Pseudomonas, Klebsiella* and *Enterobacter*. Fungi detected most commonly included 'common environmental fungi' and *Aspergillus* spp. However, despite the presence of some pathogenic species, concentrations were considered very low and unlikely to pose a significant risk.

The USA study[36] concluded that MRFs do not appear to pose a significant threat to public health or the environment, although it acknowledged that awareness of bioaerosols was increasing and this area may warrant additional evaluation. It also concluded that fugitive dusts may cause nuisance (although these could be mitigated through maintenance of roadways). It was also acknowledged that, at present, there are no occupational exposure limits for bioaerosols, making it difficult to draw conclusions concerning the health of the operatives in the MRFs.

A later study on MRFs in Denmark[4] showed metal concentrations of lead and mercury within 'normal' ranges, and only minor differences in trace metals such as iron, zinc and copper. However, cadmium levels were significantly increased in all waste-handling workers, thought attributable to batteries in the waste, but this was not thought harmful to health. Cadmium was also raised with smokers in all groups. This study concluded there might be areas of concern to MRF workers from exposure to metals.

In the UK little work had been initiated on waste handling and health effects. One study[39] reported bioaerosol concentrations to 10^5 cfu m^{-3}, and in terms of symptoms reported by workers showed that 51% reported nasal irritation, 38% throat irritation, 21% eye irritation, 38% dry cough, 31% joint pains and 38% unusual tiredness. In this study, workers who undertook peak flow recordings (lung function) showed unusual variability when changing jobs within the MRF with a small amount of individuals experiencing a sharp drop of more than 100 l min^{-1} in relation to their work.

In Finland, bioaerosols were measured at a source segregated 'resource recovery plant' (effectively a MRF).[40] The concentrations measured were attributed to open conveyors, an inefficient ventilation system and accumulation of waste in the plant. The workers in waste processing complained of upper respiratory tract and eye irritations. Most of these symptoms were reported to probably be the result of the VOC and bioaerosol exposure. It was recommended that workers were educated in hygiene to mitigate these exposures.

In the Canadian study many factors were investigated, including bioaerosols, gases, noise, EMFs and ergonomic factors.[41] Many measurements were found to be relatively low, although the authors used a guideline of 10^4 cfu m^{-3} for bioaerosols and concluded that workers could be at risk from these exposures. However, the main conclusions were that ergonomics could be of concern, particularly back pain due to standing and sorting activities.[41]

In terms of health studies, throughout Europe and the USA few epidemiological studies concerning MRFs have been identified. A German study[42] reported

[39] T. L. Gladding and P. C. Coggins, Exposure to Microorganisms and Health Effects of Working in UK Materials Recovery Facilities – A Preliminary Report, *Ann. Agric. Environ. Med.*, 1997, **4**, 137–141.

[40] H. Kiviranta, A. Tuomainen , M. Reiman, S. Laitinen, A. Nevalainen and J. Liesivuori, Exposure to Airborne Microorganisms and Volatile Organic Compounds in Different Types of Waste Handling, *Ann. Agric. Environ. Med.*, 1999, **6**, 39–44.

[41] J. Lavoie and S. Guertin, Evaluation of Health and Safety Risks in Municipal Solid Waste Recycling Plants, *J. Air Waste Manag. Assoc.*, 2001, **51**, 352–360.

[42] E. Marth, F. F. Reinthaler, D. Haas, U. Eibel, G. Feieri, I. Wendelin, S. Jelovcan and S. Barth, Waste Treatment – Health: A Longitudinal Study, *Schriftenr. ver Wasser Boden Lufthyg.*, 1999, **104**, 569–583.

research into 256 workers from manual sorting facilities (sorted or unsorted is not specified) over a period of three years concentrating on lung function and the immune system. Observations showed a decrease in lung function and an increase in total IgE (with a fluctuation of $+200\%$ to -100%).

Conclusions from the literature.   The majority of the literature concerning waste sorting has concentrated on particulates, heavy metals and bioaerosol exposure (in particular exposure to viable particles) as being the issues of concern in MRFs. Previously, collection and disposal work was mainly an outdoor activity; therefore contaminants may have been dispersed into the atmosphere. When indoor waste sorting facilities appeared in the 1970s, a number of studies were initiated to investigate the effects of sorting primarily mixed waste.[21,22] Since the 1980s[24,25] little other research had been carried out, as plants had mostly disappeared through economic circumstances. It was only when sorting was re-introduced in Denmark that further studies were instigated, and these were in direct response to observed health effects among the workers in this plant. [25–28]

A recent Danish review[11] concluded that bioaerosols were of most concern during waste sorting, where concentrations may reach up to 10^8 cfu m^{-3} and should be considered potentially harmful, and recommended plants sorting waste should be designed to prevent bioaerosol exposure. This review shows sorting of unseparated waste generated bioaerosols in the region of 10^6–10^7 cfu m^{-3} and sorting of separated waste generating bioaerosols in the region of 10^4–10^5 cfu m^{-3}. Specific activities such as manual sorting and baling, leading to greater aerosolisation, may pose higher risks for operatives. Areas where additional exposures may occur have been recognised; for instance in Germany specific guidelines have been constructed for the protection of workers in sorting plants.[18] This specifies that workers should be separated from waste and that personal protection should be supplied if separation is not possible. It was also thought that exposure might be linked to method of operation or waste inputs.

In terms of health effects, it is thought that, because of the relatively short employment time of this 'new' industry, chronic health effects are not yet being reported.[43] Symptoms most commonly seen in the research are pulmonary disorders, organic dust-like symptoms, gastrointestinal problems, eye inflammation, and irritation of the skin and upper airways. The term 'waste recycling worker syndrome' has been suggested for the fever, influenza-like symptoms, upper airway irritation and eye inflammation often seen in waste handling.[44] However, limited information exists on the magnitude of risks and the causal factors of these problems, particularly in relation to different facilities and different working tasks.

Danish research concluded that the causative agents in regard to symptoms derive from microbiological activity, and are a complex mixture of endotoxins

[43] T. Sigsgaard, A. Abel and L. Donbaek, Lung Function Changes among Recycling Workers Exposed to Organic Dusts, *Am. J. Ind. Med.*, 1994, **25**, 69–72.

[44] O. M. Poulsen, N. O. Breum, N. Ebbehøj, A. M. Hansen, U. I. Ivens, D. V. Lelieveld, P. Malmros, L. Matthiasen, B. H. Nielsen, E. M. Nielsen, B. Schibye, T. Skov, E. I. Stenbaek and C. K. Wilkins, Collection of Domestic Waste. Review of Occupational Health Problems and Their Possible Causes, *Sci. Total Environ.*, 1995, **170**, 1–19.

from Gram-negative bacteria, glucan from fungi, and possibly enterotoxin from bacteria.[28] At present, in Denmark and some other European countries, the handling of waste is associated with a range of respiratory diseases and symptoms in exposed individuals, *e.g.* dry cough, exercise-induced dyspnoea, asthma, ODTS, diarrhoea and gastrointestinal problems thought attributable to microorganisms containing a mixture of fungal spores, bacterial endotoxins and allergenic proteins with toxic and immunological influence. Early research emphasised viable microorganisms (colony forming units), which has evolved into total counts, endotoxins and lately glucan analysis. This represents a series of 'discoveries' concerning the potent aspects of organic dusts, which may develop further in the future. The exact mechanisms causing symptoms associated with waste handling and the components or combination of components of the dust that elicit specific effects are not well known. The future determination of constituents and dose–response relationships is crucial.

Considering these issues, a study was instigated in the UK to compare MRFs and the health effects experienced by MRF workers; this is outlined in brief below.

3   A Further Study on MRFs

From the literature review, it can be determined that there are various gaps in the reported studies where there is little research concerning MRFs. This was recognised both by European BIOMED2 funding, and by the Environment Agency for England and Wales. They co-funded a study to provide information on issues such as physical and chemical hazards in MRFs, and potential bioaerosol exposures in relation to actual health effects. This study was performed in eleven MRFs throughout England and Wales handling a mixture of household and commercial waste materials, with nine participating in the collection of health data. The MRFs in this study accepted materials from the householder utilising different collection systems.

Dust and bioaerosol measurements, including measurements for endotoxin and glucan, were made using methodologies reported in the previous papers.[7,8,12,13,24,31,35,39] Also measured were Volatile Organic Compounds (VOCs), cadmium and mercury, using similar methodologies to those reported in previous papers.[4,5,40] Electromagnetic fields (EMFs) were measured for the first time in MRFs in the UK. Cross-sectional questionnaires were given as a personal interview to each operative working within the nine MRFs ($n = 159$) during 1999. The questionnaire has recently been standardised and used in a research project on health effects among waste handlers in the BIOMED2 programme [www.miljomedicin.gu.se/ev-projekt/index.htm]. It is a proposed standard questionnaire for workers in the waste industry. Questions covered previous work history, type of work carried out, relevant out-of-work activities and smoking habits. These were followed by questions on symptoms specifically related to work, *e.g.* cough (dry or with phlegm), chest tightness, eye, nose and throat irritations, itchy or congested nose, nausea, diarrhoea. Episodes of fever and influenza-like symptoms were also specifically targeted. Blood (three MRFs) and lung function (two MRFs) data were also taken in a similar manner to previous studies.[12,26,39,43]

Table 6 Concentrations of dust, endotoxin and glucan

Parameter	N	Min	Max	Mean
Total dust (mg m^{-3})	260	0	62.61	6.27
Endotoxin (ng m^{-3})	128	0.19	198.17	10.89
Glucan (ng m^{-3})	119	0	137.37	18.84

Table 7 Exposure splits

Variable	Exposure groups	Site numbers	Exposure levels	No. workers per group
Total dust	Higher	4, 5, 7	>5.0 mg m^{-3}	42 (26.4%)
	Medium	2, 3, 9	3.0–4.9 mg m^{-3}	78 (49.1%)
	Lower	1, 6, 8	<3.0 mg m^{-3}	39 (24.5%)
Endotoxin	Higher	3, 4, 7	>8.0 ng m^{-3}	75 (47.2%)
	Medium	6, 8, 9	4.0–7.9 ng m^{-3}	49 (30.8%)
	Lower	1, 2, 5	<4.0 ng m^{-3}	35 (22.0%)
Glucan	Higher	3, 4, 8	>12.0 ng m^{-3}	60 (37.7%)
	Medium	1, 7	5.0–12.0 ng m^{-3}	42 (26.4%)
	Lower	2, 5, 6, 9	<5.0 ng m^{-3}	57 (35.8%)

Analysis used standard SPSS-PC software. Bivariate analyses were carried out using conventional χ-square methods, tests for trends, and appropriate non-parametric independent samples tests. Differences were considered statistically significant at $p < 0.05$. Multivariate analysis using logistic regression estimated adjusted odds ratios with 95% confidence intervals (controlling for smoking status, age and sex).

Results of note. Exposure results are summarised in Table 6.[45]

To evaluate the reported symptoms according to exposure levels within the MRFs, the mean and median of dust, endotoxin and $(1 \rightarrow 3)$-β-D-glucan exposure measurements were plotted. Sites were then ranked according to their exposure levels based on this information. The results of this can be seen in Table 7.

This split was taken by plotting mean *vs.* median exposure between the sites. There were also sufficient amounts of workers in each group in which to carry out a meaningful statistical analysis of amount of symptoms experienced. Using this information, cross-tabulations for symptoms at the difference sites were compared for the different exposure classifications and number of symptoms. Two tests for significance were carried out, one concerning whether there was a significant difference between the groups, and another to test for linear-by-linear association (to determine whether linear trends in relation to exposure are significant).

Further significance testing ascertained that current smokers were more heavily concentrated in the dust high exposure group (69% of the individuals in that group compared to 53.8% in the lower and 62.8% in the medium exposed group) and was marginally significant ($p = 0.054$), and so this should be considered. However, this was not true for endotoxin ($p = 0.964$) and was marginal for glucan ($p = 0.060$). Age, sex and presence of chronic disease such as

[45] T. L. Gladding, J. Thorn and D. Stott, Organic Dust Exposure and Work-Related Effects among Recycling Workers, *Am. J. Ind. Med.* (submitted).

Table 8 Exposure *vs.* No. of workers suffering named symptom (in %)

Symptom/Exposure	Higher exp.	Middle exp.	Lower exp.	Linear	χ
Total dust					
Itchy red skin	4.8	20.5	7.7	0.648	0.026*
Skin rash	2.4	10.3	0	0.691	0.043*
Irritated nose/sneezing	66.7	50.0	46.2	0.062	0.125
Diarrhoea	45.2	44.9	20.5	0.026*	0.024*
Flu symptoms	21.4	42.3	13.2	0.493	0.002*
Endotoxin					
Cough with phlegm	33.3	20.4	17.1	0.048*	0.114
Dry cough	37.3	16.3	40.0	0.760	0.022*
Stuffy nose	64.0	61.2	45.7	0.090	0.180
Hoarse/parched throat	33.3	12.2	20.0	0.048*	0.022*
Glucan					
Cough with phlegm	33.3	28.6	16.7	0.039*	0.107
Hoarse/parched throat	31.6	28.6	13.3	0.021*	0.049*
Chest tightness	15.8	11.9	5.0	0.059	0.161
Stuffy nose	66.7	57.1	53.3	0.145	0.326
Irritated nose/sneezing	47.4	66.7	50.0	0.796	0.130

*Indicates significant association between exposure and health ($p = 0.05$)

asthma did not show any differences between exposure groups. Length of time working at a MRF may be unevenly distributed between exposure groups, with a bias towards longer serving workers being in the higher exposure groups. When significance tests carried out for high *vs.* low exposure to dust and endotoxin no significant differences were found between exposure, duration of working or job. However, with glucan using χ-square it appears that workers in the higher exposed bracket have been working at MRFs longer, a difference found to be significant ($p = 0.032$). This could have a potential confounding effect on results, but could also be interpreted as a 'healthy worker selection' effect. Previous studies have reported under-representation of asthma among waste handling workers.[38,46] In particular, a recent German study found that individuals with atopic diseases were significantly under-represented in compost workers ($p = 0.003$) indicating a healthy worker selection.[46] Therefore it is possible that health effects due to exposure to bioaerosols are likely to be underestimated in waste management. Table 8 illustrates symptoms compared to exposure grouping. Gradients can clearly be seen between those sites classified as higher compared to lower exposed sites. Adjusted odds ratios (for smoking, age and sex) were then used to compare low exposure to medium, and low to high exposure in relation to reported symptoms. The purpose of this exercise was to detect any dose–response relationships, where an increased odds ratio would be visible from low to medium compared to low to high exposure. The results are seen in Table 9.

These results indicate that, for the first time, exposure to dust, endotoxin and

[46] J. Bünger, M. Antlauf-Lammers, T. G. Schulz, G. A. Westphal, M. M. Müller, P. Ruhnau and E. Hallier, Health Complaints and Immunological Markers of Exposure to Bioaerosols among Biowaste Collectors and Compost Workers, *Occup. Environ. Med.*, 2000, **57**, 458–464.

Table 9 Adjusted odds ratios for symptoms *vs.* exposure

Symptom	Low vs. medium		Low vs. high	
Total dust				
Irritated nose/sneeze	1.0819	(0.4848–2.4140)	2.6869	(1.0476–6.8914)*
Diarrhoea	3.4162	(1.3761–8.4807)*	3.5559	(1.2945–9.7676)*
Flu symptoms	3.6438	(1.3496–9.8383)*	1.3651	(0.4278–4.3559)
Endotoxin				
Cough with phlegm	1.2758	(0.4112–3.9586)	2.7082	(0.9724–7.5424)
Stuffy nose	1.9825	(0.8061–4.8760)	2.3572	(1.0094–5.5042)*
Glucan				
Chest tightness	3.5350	(0.7480–16.7059)	5.2799	(1.2653–22.0322)*
Cough with phlegm	2.2844	(0.8578–6.0836)	2.6736	(1.0829–6.6007)*
Hoarse/parched throat	2.4026	(0.8699–6.6357)	3.5217	(1.3430–9.2350)*
Stomach problems	2.3113	(0.4752–11.2422)	5.7389	(1.4465–22.7692)*
Irritated nose/sneeze	2.4125	(1.0246–5.6805)*	0.8899	(0.4133–1.9162)
Stuffy nose	1.2867	(0.5630–2.9408)	2.3138	(1.0333–5.1810)*

*Indicates significant association between exposure and health ($p = 0.05$)

glucan in a MRF environment shows a dose–response relationship in terms of exposure and health, particularly for respiratory and gastrointestinal effects. These results illustrate that total dust exposures are related to diarrhoea and skin problems mainly, although upper respiratory nose and throat irritations are also apparent. The situation with endotoxin is more complex. It appears those in medium-exposed MRFs suffer the least amount of symptoms to a significant level for those illustrated in Table 9. Workers exposed to higher levels of glucan may be more prone to developing a range of health symptoms. The symptoms reported here are not unusual in workers in the waste industry. In general, increased exposure to $(1{\rightarrow}3)$-β-D-glucan has been associated with symptoms such as airway inflammation, fatigue and headache.[8]

However, it should be borne in mind that the questionnaire was based on self-reporting of certain symptoms. The workers may be more prone to report their symptoms if the symptoms have been discussed in association with working in the MRF. Thus, there is a possibility for over-reporting. However, when workers responded to the questionnaire they were not aware of the exposure levels.

However, taking these issues into consideration, in common with other studies in the waste industry a significant association has been demonstrated in this study between exposure and symptoms related to endotoxin, glucan and upper-respiratory disorders, systemic effects and gastrointestinal effects.[12,26,28,38,43] In particular, symptoms are reported in MRFs with a higher exposure to dusts, most commonly the twin-wheeled bin facilities. These types of facilities typically have a higher amount of reject materials entering the plants than box or bag type MRFs. In the two twin-wheeled bin MRFs studied, reject rates in the region of 40% were reported, compared to less than 10% for all other types of MRF. This study has found that air quality, presence of residues and worker symptoms are related.

Concerning the blood data in the MRFs with higher $(1{\rightarrow}3)$-β-D-glucan exposure a significantly decreased number of monocytes were found. The decrease in monocyte numbers in blood can reflect a recruitment from the blood to the lung among the workers exposed to $(1\rightarrow3)$-β-D-glucan. In addition, ESR was decreased among the workers exposed to higher amounts of $(1\rightarrow3)$-β-D-glucan. These results can indicate that $(1\rightarrow3)$-β-D-glucan have a blocking effect on the inflammatory response in blood. The implications of this are fully discussed in a related paper.[45]

Also measured were Volatile Organic Compounds (VOCs), electromagnetic fields, total and viable microorganisms, cadmium and mercury. However, in common with similar studies[41] these results did not show any significant amounts in the MRFs. Lead was detected in the air of one facility, and was found in settled dust in all of the MRFs measured in very small amounts (data not shown). It is not expected that these parameters significantly affected MRF workers.

In conclusion, the results suggest that MRF workers exposed to higher levels of total dust, endotoxin and $(1\rightarrow3)$-β-D-glucan at their work sites exhibit various work-related symptoms, primarily respiratory and gastrointestinal in origin.

Future Research

Further research is required in this important area. For instance, interaction between airborne exposures (bioaerosols and diesel particles and/or chemicals such as VOCs) is a subject for future research in waste facilities, particularly focusing on the inflammatory responses often seen in this occupation.[12,26,28,38,43,45] Indeed, a recent study demonstrated that dust generated from mixed household waste handling had, in itself, a high inflammatory potential.[47] Further work is needed to determine the constituents of this dust; endotoxin was not linked to the inflammatory potential of this dust.[47] Working outdoors in close proximity to vehicles may expose workers to diesel exhaust particles (DEP), or DEP may be drawn into vehicle cabs through open windows, doors or inefficient cab filters. DEP is known to cause irritation of the upper respiratory tract[48] and it has been recognised that this may contribute to health risks for waste handlers, both on its own and as an immunostimulant.[11] Recent *in vitro* and animal model studies[49,50] have indicated that there may be an adjuvant effect of DEP on pulmonary inflammation response caused by exposure to endotoxin. As endotoxin and DEP may both be present together in the air in the vicinity of waste handling operations, a combined effect from the two respiratory insults may be possible.

4 Conclusions

The health risks associated with MRFs are slowly being elucidated, and they need

[47] L. Allermann and O. M. Poulsen, Inflammatory Potential of Dust from Waste Handling Facilities Measured as IL-8 Secretion from Lung Epithelial Cells *In Vitro*, *Ann. Occup. Hyg.*, 2000, **44** (4), 259–269.

[48] P. T. J. Scheepers and R. P. Bos, Combustion of Diesel Fuel from a Toxicological Perspective, II Toxicity, *Int. Arch. Occup. Environ Health*, 1992, **64**, 163–177.

[49] W. Dong, J. Lewtas and M. I. Luster, Role of Endotoxin in Tumor Necrosis Factor alpha Expression from Alveolar Macrophages Treated with Urban Air Particles, *Exp. Lung Res.*, 1996, **22**, 577–592.

[50] H. M. Yang, M. W. Barger, V. Castranova, J. K. Ma, J. J. Yang and J. Y. Ma, Effects of Diesel Exhaust Particles (DEP), Carbon Black, and Silica on Macrophage Responses to Lipopolysaccharide: Evidence of DEP Suppression of Macrophage Activity, *J. Toxicol. Environ. Health*, 1999, **58**, 261–278.

further evaluation. The Danish Working Environment Service has banned handsorting of mixed domestic waste.[27] The Danish have recognised that there are health problems associated with collection and recycling of household waste. Undoubtedly these are exacerbated by the behaviour and personal hygiene of operatives, and working environment guidelines were initiated to mitigate these circumstances.[27] Other types of waste sorted manually must be kept free of moisture and contamination with household waste. Similar approaches should be considered in the UK.

It becoming apparent that handsorting of recycled materials carries with it health issues which must be addressed. The difficulty arises as income from a MRF will never equal that from a landfill or incinerator. Many of the materials collected are low value. Conversely, initial investment in a MRF, which could be as low as several thousand pounds, will be significantly less than for other waste management options (requiring investment running into the millions). For this reason alone MRFs may become an even more popular waste management option in attempting to comply with the requirements of the EU landfill directive. This means a dramatic increase in the number of workers handsorting waste materials. In view of the research thus far, the health risks to these workers and possible mitigation factors should be an important consideration when planning this strategy.

5 Acknowledgements

Grant sponsor: The European Commission and the Environment Agency, Bristol, England
Grant number: Contract BMH4-CT96-0105 and R&D Project P1-214
Dr David Stott of the University of Luton for statistical advice, Dr Jörgen Thorn of the University of Gothenburg for medical input.

Microbial Emissions from Composting Sites

JILLIAN R. M. SWAN, BRIAN CROOK
AND E. JANE GILBERT

1 Introduction

The composting process can be defined as: the controlled biological decomposition and stabilisation of organic substrates, under conditions that are predominantly aerobic and that allow the development of thermophilic temperatures as a result of biologically produced heat. It results in a final product that has been sanitised and stabilised, is high in humic substances and can be beneficially applied to land, which is typically referred to as 'compost' (modified from Haug[1]).

Farmers have practised composting for centuries, returning waste organic residues to the land in order to maintain soil fertility and organic matter. It is only relatively recently that modern society has begun to recognise the important role composting has to play in managing the ever increasing quantities of waste it produces. Composting is now employed as a treatment process for a wide range of organic substrates such as municipal solid wastes, sewage sludges, and agricultural and industrial by-products.[2] Actively composting materials, or finished composts, have been shown to degrade a wide range of organic pollutants, and are thus used in the bioremediation of contaminated soils.[3]

There are a number of ways in which composting can be carried out, ranging from small-scale composting of garden wastes by householders, through medium-scale composting by community groups and farmers, to large-scale composting at specially designed sites, although the range of technologies at the latter can vary considerably.

Composting Practices

The activities carried out at a composting site, irrespective of its size, are aimed at amending or controlling feedstock structure, composition, temperature or oxygen content.

[1] R. T. Haug, *The Practical Handbook of Compost Engineering*, Lewis, Boca Raton, Florida, 1993.
[2] E. J. Gilbert, D. S. Riggle and F. D. Holland, *Large-scale Composting – A Practical Manual for the UK*, The Composting Association, Wellingborough, UK, 2001.
[3] K. T. Semple, B. J. Reid and T. R. Fermor, *Environ. Pollut.*, 2001, **112**, 269–283.

Issues in Environmental Science and Technology, No. 18
Environmental and Health Impact of Solid Waste Management Activities
© The Royal Society of Chemistry, 2002

Most materials received at a composting facility require processing prior to composting. In practice, there are four main activities, namely: shredding, to reduce particle size and increase the surface area to volume ratio; mixing different feedstocks together to improve homogeneity and adjust the carbon to nitrogen (C:N) ratio and/or moisture content; adding water where particularly dry materials are received; and removing contaminants.

The degree of process control employed during the active composting phase varies depending upon the size and location of the site, the nature of the feedstocks and the intended end uses of the composted materials. In practice there are four principal approaches that can be adopted for composting wastes on a large-scale.

Windrow systems. Open-air turned windrows are the simplest system, and they are widely used to compost green wastes (sometimes called botanical residues or yard trimmings). Feedstocks are laid out in long piles called 'windrows', usually shaped as an elongated triangular prism, although the exact shape varies according to the material and equipment used. The term originates from the farming practice of piling hay in rows so that it will dry out in the wind.

These windrows are 'turned' to blend the composting mix, introduce fresh air, and release trapped heat, moisture and stale air. In practice, the technique often involves breaking up the windrows by lifting the composting materials into the air and allowing them to drop back down.

At smaller facilities, a tractor with a front-end loader or grab is often used to scoop up portions of the composting materials. Each bucket-load is carried to another area and emptied onto the ground, re-forming a new windrow. At larger sites, which process greater quantities of material, specialised windrow turning machines are often employed.

Aerated static piles. Some mechanised systems may dispense with turning, and either blow (positive aeration) or suck (negative aeration) air through the composting materials. These 'forced aeration' systems often rely on a perforated pipe running through the pile or a trough underneath the composting materials, which is attached to an air compressor. However, some operators may combine both approaches, using a forced aeration system with some element of physical turning.

The rate of aeration may be linked to oxygen concentration and/or temperature *via* a negative feedback system, or fans may be switched on and off for defined periods of time. The exhaust gases from negative pressure systems may be passed through a biofilter or scrubber before discharge to the environment, whereas in positive pressure systems the piles may be covered with layers of mature compost to act as an *in situ* biofilter. These techniques have been widely employed in the USA to compost sewage sludges.

In-vessel systems. Unlike windrows and aerated static piles, in-vessel systems contain the composting feedstocks in vessels that are usually enclosed, which affords a much greater level of process and emission control. Wide ranges of systems are marketed, each with their own unique benefits and applicability to different feedstocks and situations. In broad terms there are six different types of in-vessel system, although many may be classified into more than one category:

- Containers – These are generally small-scale systems designed for decentralised use, especially food processing and catering wastes. Most systems operate as batch units, supplying air through perforations in the floor of the container.
- Tunnels – These tend to be larger and more sophisticated than containers, and have been adapted from the mushroom composting industry. They operate in similar ways to containers although some element of mechanical agitation may be employed. Both batch and continuous systems have been developed.
- Agitated bays – These consist of rows of rectangular beds separated by low walls on each side along which turning and shredding machines run. The machines mix the compost and deposit it further along the bay in a continuous flow; forced aeration may also be provided via flooring ducts.
- Rotating drums – These are large rotating cylinders (generally 3–4 m in diameter, and anywhere up to 50 m long) that are slightly inclined from the horizontal. Feedstocks are introduced in the top end and mixed and fragmented as they move towards the outlet. Water and nutrients may be added and forced aeration provided.
- Silos or tower systems – These are vertical units that operate on a continuous basis. Feedstocks are loaded into the top of the unit and are composted as they pass down through the unit. The resultant composts are harvested at the bottom of the vessel using augers.
- Enclosed halls – These compost material on the floor of the hall and are usually contained in one long bed. The whole composting process tends to occur in the same hall; large bucket wheels are used to turn and move the material through the system.

Vermicomposting.   Vermicomposting is the process of using selected species of earthworms to help compost organic wastes, and stems from the established business of vermiculture (the breeding of earthworms, mainly for the fishing bait market). It is usually carried out in long troughs in which the temperature is kept below 35 °C. With traditional composting, the compost piles are mixed and aerated mechanically, whilst with vermicomposting it is the earthworms that fragment, mix and help aerate the waste.

Post processing.   Following composting, most facilities grade the composts into different particle size fractions to create products for varying end uses, and to remove contaminants or partially composted fragments. This process is termed 'screening' and often involves the use of purpose-built machines, of which there are three common types, namely oscillating or shaker screens, rotary trommel screens, and disc or star screens.

Screening compost is the most common method of adding value. Most of the compost sold as soil improvers in the UK is screened to a diameter of 10 mm or less, most mulches to a diameter of 11–25 mm, although different markets and intended uses require different particle sizes. Blending composts with other

Table 1 Key stages of the composting process

Stage	Key features	Stage characteristics	Approximate duration
High rate composting	Micro-organisms consume forms of carbon they can easily break down, *e.g.* sugars and starches	High rate of biological activity characterised by high oxygen demand and of heat generation rates Tendency for pH to initially drop below the optimum of 6–8, then rise above 8 as composting proceeds	4–40 days depending upon system type
Stabilisation	Micro-organisms consume forms of carbon they can break down fairly readily, *e.g.* cellulose	Biological activity starts to decline. Oxygen demand gradually decreases. Declining heat generation Tendency for pH to remain above 8	20–60 days depending on system type
Maturation (Curing)	Amount of available carbon is much reduced and microbial consumption slowed down. Re-colonisation by soil microbes	Reduced biological activity. Medium to low oxygen demand Little heat generation; temperature should be below 50 °C Oxidisation of ammonium to nitrate ions Tendency for pH to fall towards neutral (7)	Variable duration depending upon test method used and intended end use

Source: Gilbert *et al.*[2]

materials (such as coir and/or artificial plant nutrients) is often carried out to create high-value products for specialist uses in horticulture and turf management.

The Composting Process

In very simple terms there are three key stages to the composting process (Table 1), although they are by no means mutually exclusive, and are dependent upon the feedstocks and processing conditions employed.

The biochemistry and microbiology of composting remain poorly understood to date. Despite extensive research over the past twenty years into engineering aspects and the benefits of using composts, composting is still essentially considered a 'black box' process. This stems, in part, from the inherent complexity of the composting process, which is heterogeneous in nature and is directly influenced by factors such as feedstock composition and structure, temperature, pH, moisture, oxygen and ammonia concentrations.[4] In many cases indirect methods, such as calorimetric analyses for example, have been used to measure microbial metabolic activity.[5]

Composting relies upon the inter-related activities of a diverse range of micro-organisms to convert organic waste substrates into a stabilised material ('compost'), which is high in humic substances ('humus') and contains useful plant nutrients. In most feedstocks, the principal source of carbon and energy is derived from lignocelluloses.[6] Cellulase activities in composting materials have been widely studied and correlated to decreases in cellulose content.[7] The degradation of recalcitrant lignins in composting systems has been less well characterised, although thermophilic microfungi, and to a lesser extent actinomycetes, are thought to play key roles.[8]

Humification (the process of forming humus) is complex and thought to involve a number of degradative and condensation reactions involving lignins, carbohydrates and nitrogenous compounds.[6,8] Nuclear magnetic resonance spectroscopy, gas chromatography–mass spectrometry and Fourier transform infrared spectroscopy have all been used to track changes in feedstock composition and the formation of humic substances (for example González-Vila, *et al.*[9]).

The composting process can be split into three key stages based on changes in temperature:[4]

- Phase 1 is characterised by an increase in temperature from ambient as a result of microbial metabolic activity and has been termed the 'high rate' composting phase. During this phase simple carbohydrates and proteins are readily degraded, firstly by mesophiles, which are then succeeded by thermotolerant and thermophilic species as the temperature rises above 45 °C.
- Phase 2 has been termed the 'stabilisation' phase and is characterised by the attainment of thermophilic temperatures (> 50 °C), which selects for thermophilic bacteria.[10] However, this may be an over simplistic assumption, as the survival of isolates typically characterised as mesophiles has been suggested by Droffner *et al.*[11] The thermophilic stage plays a key role in the thermal destruction of pathogenic micro-organisms, weed seeds and propagules, although antagonisms such as competition and the formation of secondary metabolites may be significant.[12]

The thermophilic composting phase has received the greatest attention to date,

[4] F. C. Miller, in *Microbiology of Solid Waste*, ed. A. C. Palmisano and M. A. Barlaz, CRC Press, Boca Raton, Florida, 1996.

[5] P. Weppen, *Biomass Bioenerg.*, 2001, **21**, 289–299.

[6] J. M. Lynch, in *Science and Engineering of Composting: Design, Environmental, Microbiological and Utilization Aspects*, ed. H. A. J. Hoitink and H. M. Keener, Renaissance Publications, Worthington, Ohio, 1993, pp. 24–35.

[7] F. J. Stutzenberger, A. J. Kaufman and R. D. Lossin, *Can. J. Microbiol.*, 1969, **16**, 553–560.

[8] M. Tuomela, M. Vikman, A. Hatakka and M. Itävaara, *Bioresource Technol.*, 2000, **72**, 169–183.

[9] F. J. González-Vila, G. Almendros and F. Madrid, *Sci. Total Environ.*, 1999, **236**, 215–229.

[10] P. F. Strom, *Appl. Environ. Microbiol.*, 1985, **50**, 899–905.

[11] M. L. Droffner, W. F. Brinton, Jr. and E. Evans, *Biomass Bioenergy*, 1995, **8**, 191–195.

[12] J. Sidhu, R. A. Gibbs, G. E. Ho and I. Unkovich, *Water Res.*, 2001, **35**, 913–920.

especially composts produced for the cultivation of mushrooms on a commercial scale.[13] Thermophilic actinomycetes,[14] *Bacillus* species[13,15,16] and *Thermus* species[17] have all be shown to dominate, whilst thermotolerant fungi from the genera *Aspergillus* and *Penicillium* have been widely reported.[4,18]

However, species diversity is thought to decrease at high temperatures,[10,15] whilst Gram-positive bacteria have been shown to predominate.[15,19]

• Phase 3 is the 'maturation' phase and is typically characterised by a reduction in temperature towards ambient as a result of decreases in metabolic activity following oxidation of readily biodegradable substrates. Mesophilic actinomycetes and fungi begin to predominate during this stage, and are thought to be responsible for degrading and converting lignins, which occurs optimally at these lower temperatures.[8]

In recent years a number of approaches have been adopted to monitor the changes in microbial communities during the composting process. Dynamic changes have been tracked using carbon source utilisation and phospholipid fatty acid analyses.[20-22] Advances in nucleic acid techniques are now beginning to shed light on the roles of microbial communities during the composting process.[23] Many of these techniques have been applied in soil microbiological studies, where the phenomenon of viable but not culturable states is known to exist.[24] Consequently genera previously not identified in composts using classical plate culture techniques are now being identified (see for example, refs. 17 and 19). Blanc *et al.*[23] described the characterisation of Operational Taxonomic Units based on endonuclease restriction profiles of cloned 16S ribosomal DNA recombinants isolated from hot composts, and measured changes in population diversity in young and old composts. Similarly, Peters *et al.*[25] demonstrated changes in microbial communities at different stages of the composting process using PCR amplification of small-subunit ribosomal RNA genes.

2 Bioaerosol Components

For the reasons described above, high concentrations of bacteria and fungi are present in composts. For example, Dees and Ghiorse[19] reported total counts in the order of 10^{10} cells per gram of compost [dry weight] measured using epifluorescence microscopy, with thermophilic heterotrophic aerobes measured

[13] J. G. Kleyn and T. F. Wetzler, *Can. J. Microbiol.*, 1981, **27**, 748–753.

[14] J. Lacey, *Ann. Agric. Environ. Med.*, 1997, **4**, 113–121.

[15] A. Ghazifard, R. Kasra-Kermanshahi and Z. E. Far, *Waste Manage. Res.*, 2001, **19**, 257–261.

[16] P. F. Strom, *Appl. Environ. Microbiol.*, 1985, **50**, 906–913.

[17] T. Beffa, M. Blanc, P.-F. Lyon, G. Voght, M. Marchiani, J. Lott Fischer and M. Aragno, *Appl. Environ. Microbiol.*, 1996, **62**, 1723–1727.

[18] P. D. Millner, P. B. Marsh, R. B. Snowden and J. F. Parr, *Appl. Environ. Microbiol.*, 1977, **34**, 765–772.

[19] P. M. Dees and W. C. Ghiorse, *FEMS Microbiol. Ecol.*, 2001, **35**, 207–216.

[20] R. F. Herrmann and J. F. Shann, *Microb. Ecol.*, 1997, **33**, 78–85.

[21] B. Hellmann, L. Zelles, A. Palojärvi and Q. Bai, *Appl. Environ. Microbiol.*, 1997, **63**, 1011–1018.

[22] L. Carpenter-Boggs, A. C. Kennedy and J. P. Reganold, *Appl. Environ. Microbiol.*, 1998, **64**, 4062–4064.

[23] M. Blanc, L. Marilley, T. Beffa, and M. Aragno, *FEMS Microbiol. Ecol.*, 1999, **28**, 141–149.

[24] J. Kondrój and J. D. van Elsas, *J. Microbiol. Methods*, 2001, **43**, 197–212.

[25] S. Peters, S. Koschinsky, F. Schweiger and C. C. Tebbe, *Appl. Environ. Microbiol.*, 2000, **66**, 930–936.

in the order of 10^8 colony forming units (cfu) (g dry wt)$^{-1}$. Lacey[14] reported actinomycetes in mushroom composts in the order of 10^6 cfu (g dry wt)$^{-1}$; Strom[10] and Millner *et al.*[18] reported concentrations of *Aspergillus fumigatus* in excess of 10^5 cfu (g dry wt)$^{-1}$ in composting sewage sludge, whilst Beffa *et al.*[17] reported concentrations of thermophilic bacteria related to the genus *Thermus* in the range of $10^7–10^{10}$ cells (g dry wt)$^{-1}$. A cfu is defined as the unit of one or more cells or spores which when inoculated onto suitable growth medium grows to form a single colony.

Whenever composting materials are moved around, for example during the shredding, turning and screening processes, these micro-organisms can be aerosolised, forming what is termed a bioaerosol. Actively managed, medium- to large-scale composting harnesses the activity of indigenous micro-organisms commonly present in the soil that naturally decay, such as fallen leaves. Therefore, the microbial components of bioaerosols generated during the composting process contain many of the same micro-organisms that are commonly isolated from 'normal' outdoor air. The main difference is that of scale. The handling of large quantities of compost potentially can lead to the release into the air of large quantities of the bacteria, fungi and actinomycetes and their components, found in compost, as a bioaerosol.

Exposure to the micro-organisms found in compost could potentially cause ill-health in the people exposed to them either by infection, allergy or an adverse response to toxins. The composting process generates heat, so any human pathogens present in the raw materials, such as coliform bacteria from faecal material which could give rise to gastro-intestinal infection, should be rapidly killed off during the composting process.

Some of the micro-organisms which increase in number during the composting process are toxic and/or allergenic and still have the potential to cause problems when they are dead.

There are two main routes of exposure to compost micro-organisms: ingestion of the micro-organisms or inhalation of bioaerosols created during the handling of compost. Good hygiene practices such as wearing of gloves and provision of hand washing facilities on composting sites should control risks from ingestion. However, control of bioaerosols generated by composting processes is more complex.

In order to understand the potential health hazards associated with exposure to compost bioaerosols, it is first important to examine in detail the microbial components of bioaerosols generated during the handling of compost.

Fungi

During the handling of fresh green waste the micro-organisms present are predominantly the saprophytic 'field' fungi such as *Cladosporium* spp., *Alternaria* spp., and *Verticillium* that colonise plants during growth. As the composting process progresses, the numbers and types of associated contaminants change. Some fungal spores naturally present in low numbers may germinate and grow. These are referred to as 'storage fungi' because they flourish in stored organic materials. They include *Aspergillus, Eurotium, Penicillium, Trichoderma, Absidia,*

Mucor and *Rhizopus* species. These fungi can grow at lower water availability and predominate over the field fungi, which are less well suited to multiplication in these conditions. The increased metabolic activity can lead to spontaneous heating of the compost. This can result in a succession of microbial growths so that, if enough water is available, temperatures of 65 °C can be reached, allowing the growth of thermophilic and thermotolerant fungi.[26–29]

A. fumigatus is particularly important in the composting process due to its capacity to degrade cellulose and hemicelluloses. Its optimum growth temperature is 37 °C and good growth can occur between 30 and 52 °C. Consequently it is likely to be present in significant numbers. However, its presence is also an important consideration from a human health viewpoint. It is an allergenic fungus and is an opportunistic pathogen which can cause aspergillosis in immunocompromised subjects.[30,31] *A. fumigatus* can also produce mycotoxins which are toxic, carcinogenic, and mutagenic. Other *Aspergillus* species, and other fungal species present in compost, such as *Rhizopus*, are also allergenic.

Bacteria

Bacteria can be divided into two main types: Gram-negative bacteria and Gram-positive bacteria. Gram-negative bacteria predominate in dusts of plant origin: bacteria such as *Pseudomonas* spp., *Klebsiella* spp., *Pantoea agglomerans*, *Rahnella* spp. and *Alcaligenes* spp. are commonly present.[26] Gram-positive bacteria predominate in dusts of animal origin, but are also present in dusts of plant origin, bacteria such as *Corynebacteria*, *Bacillus* spp. and cocci such as *Staphylococcus* spp. *Micrococcus* spp. and *Streptococcus* spp.[26] Gram-negative bacteria found in in dusts of animal origin include *Acinetobacter* and *Enterobacter* spp. Consequently, these species will be present in composts. Other Gram-negative bacteria of animal faecal origin, the coliform bacteria such as *E. coli*, *Campylobacter* and *Salmonella* species, could be present in compost feedstock depending on its origin. Composts prepared from animal manures obviously are most likely to contain coliforms, but organic waste may be contaminated by animal faeces.

The elevated temperatures achieved in composting should kill off coliform bacteria (see also details in Section 4, Case Studies), although it should be recognised that inadequate compost turning could lead to temperature stratification and survival of coliforms in cooler layers. This is particularly of concern with highly pathogenic strains of *E. coli*, such as the verocytotoxic *E. coli* O157, which has been shown to survive for several days in soils.[32,33] Workers handling composts from animal manures need to take additional hygiene precautions, such as thorough hand washing, and public access to composting animal

[26] J. Dutkiewicz, *Ann. Agric. Environ. Med.*, 1997, **4**, 6–11, 11–16.
[27] J. Lacey, in *Occupational Pulmonary Disease: Focus on Grain Dust and Health*, ed. J. A. Dosman and D. J. Cotton, 1980, pp. 189–200.
[28] C. S. Darke, J. Knowelden, J. Lacey and A. Milford-Ward, *Thorax*, 1976, **31**, 294–302.
[29] J. Lacey and B. Crook, *Ann. Occup. Hyg.*, 1988, **32**, 513–533.
[30] H. Allmers and X. Baur, *Am. J. Ind. Med.*, 2000, **37**, 438–442.
[31] W. Vincken, *Thorax*, 1984, **39**, 74–75.
[32] A. Maule, *Symp. Ser. Soc. Appl. Microbiol.*, 2000, **29**, 71S–78S.
[33] L. D. Ogden, D. R. Fenlon, A. J. Vinten and D. Lewis, *Int. J. Food Microbiol.*, 2001, **66**, 111–117.

manures, such as on open farms, should be restricted. Other bacterial pathogens of animal origin include *Leptospira*, the causative agent of Weil's disease. This bacterium multiplies in the kidneys of infected rats and is spread in contaminated urine, causing infection in humans through entry *via* skin abrasions and mucus membranes. For this reason among others, vermin control on compost sites is important.

Bacteria, therefore, can be hazardous to health as pathogens, *e.g.* coliforms as described above, but the main route of infection is by ingestion. Respiratory infection caused by bacteria is unlikely to be a significant hazard in composting. However, bacteria present in airborne dust from composts could cause allergies and may be toxin producers (see endotoxin).

Actinomycetes

In addition to bacteria as described above, actinomycetes are also found in these environments. Actinomycetes are filamentous Gram-positive bacteria that are commonly found associated with soil and plant materials. Thermophilic actinomycetes, with a growth temperature range of 30–60 °C, thrive in wet compost that has begun the self heating process. Therefore they can be used as indicator organisms for self heating of organic material, and as indicator organisms for the presence of bioaerosols generated from compost.[26,27,29] The most common species present are *Saccharopolyspora* (*Faenia*) *rectivirgula*, *Saccharomonospora* spp. including *S. viridis*, *Thermoactinomyces thalpophilus*, *Thermoactinomyces vulgaris* and *Thermomonospora* spp. Mesophilic species such as *Streptomyces* are also commonly present in high numbers.

Thermophilic actinomycete species are recognised respiratory allergens. Actinomycetes produce thousands of very small spores (1–3 μm diameter) which easily become airborne in large numbers when heavily colonised material is disturbed. Their small size means that they are potentially capable of penetrating deep into the human lung. They are primarily responsible for occupational allergic diseases such as Farmers' Lung Disease and Mushroom Workers' Lung Disease, which are forms of extrinsic allergic alveolitis.

Endotoxin

Endotoxin is found in the outer layer of the cell walls of all Gram-negative bacteria and some blue–green algae. Endotoxin is the fragments of the bacterial cell wall that contain lipopolysaccharide (LPS) as well as other compounds naturally occurring in the cell wall.[34] Gram-negative bacteria are present in the oral cavities and intestinal tracts of humans and animals; they also live on the surfaces of animals and plants. Consequently the general population is exposed to low levels of endotoxin and it is found in house dust.

Endotoxin is present in occupational settings, mainly as a component of organic dusts such as those of vegetable origin contaminated with Gram-negative bacteria (*e.g.* grain and cotton dust), dusts containing animal faeces *e.g.* in swine

[34] R. R. Jacobs, D. Heederik, J. Douwes and U. Zähringer, *Int. J. Occup. Environ. Health*, 1997, **3**, S6–S7.

J. R. M. Swan, B. Crook and E. J. Gilbert

Table 2 Summary of airborne occupational endotoxin exposure

Industry/activity	Mean airborne endotoxin concentration (ng m^{-3})	Reference
Waste water treatment	0.6–310	35
Waste handling	0–17.4[E]	36–38
Agriculture	0–770 000	39
Cotton industry	66–6936	40
Biotechnology	0.33–162.8	41
Machining MWF	0.1–767[PE]	39
Fibreglass wash water	0.4–27 800	42
Offices	0.018–1200	43

[E] Nanograms calculated from Endotoxin Units by dividing by 10.
[P] Personal sampling.

confinement buildings and poultry houses) and sewage sludge. Endotoxin is also present in compost as a result of the presence of Gram-negative bacteria and therefore is made airborne in dust from compost handling. The amount of airborne endotoxin in different occupational environments varies widely. Some typical examples are summarised in Table 2.

Inhalation of endotoxin causes both short term illness (flu-like symptoms, fever, myalgia, and malaise, *e.g.* organic dust toxic syndrome) and long term illness (*e.g.* chronic bronchitis, chronic obstructive pulmonary disease and long term decline in lung function). The acute clinical symptom response to endotoxin inhalation occurs 6–12 hours after exposure and lasts about 4 hours. Chronic exposure to endotoxin has been linked to work related symptoms and chronic decreases in lung function such as chronic inflammation leading to chronic bronchitis and reduced lung function.

Mycotoxins

Mycotoxins are non-volatile low molecular weight toxic secondary metabolites produced by fungi. Mycotoxins can be carcinogenic, neurotoxic and teratogenic. The most common route of exposure is ingestion. The toxins can cause acute or chronic disease in vertebrate animals,[44] and may contribute to occupational lung disease in workers exposed to organic dusts. *Aspergillus* spp., including *A.*

35 J. Liesivuori, M. Kotimaa, S. Laitinen, K. Louhelainen, J. Ponni, R. Sarantila and K. Husman, *Am. J. Ind. Med.*, 1994, **25**, 123–124.
36 K. Heldal and M. Bergum, *Ann. Agric. Environ. Med.*, 1997, **4**, 45–51.
37 T. Sigsgaard, P. Malmros, L. Nersting and C. Pedersen, *Am. Rev. Resp. Dis.*, 1994, **149**, 1407–1412.
38 T. Sigsgaard, J. C. Hansen and P. Malmros, *Ann. Agric. Environ. Med.*, 1997, **4**, 107–112.
39 Health and Safety Laboratory, J. R. M. Swan, 1999, unpublished data.
40 J. C. G. Simpson, R. McL. Niven, C. A. C. Pickering, L. A. Oldham, A. M. Fletcher and H. C. Francis, *Ann. Occup. Hyg.*, 1999, **43**, 107–115.
41 R. B. Palchak, R. Cohen, M. Ainslie and C. LaxHoerner, *Am. Ind. Hyg. Assoc. J.*, 1988, **8**, 420–421.
42 D. K. Milton, J. Amsel, C. E. Reed, P. L. Enright, L. R. Brown, G. L. Aughenbaugh and P. R. Morey, *Am. J. Ind. Med.*, 1995, **28**, 469–488.
43 E. Kateman, D. Heederik, T. M. Pal, M. Smeets, T. Smid and M. Spitteler, *Scand. J. Work Environ. Health*, 1990, **16**, 428–433.
44 S. Gravesen, J. C. Frisvad and R. A. Samson, *Microfungi*, Munksgaard, Copenhagen, 1994.

fumigatus, and *Penicillium* spp. produce mycotoxins and both are usually present in the dust generated during the handling of compost. Their possible role in causing respiratory symptoms is not fully understood and the presence of mycotoxins in compost dust has not been widely studied. Fischer *et al.*[45,46] have investigated the presence of secondary metabolites associated with fungi, in particular *A. fumigatus* mycotoxins, in bioaerosols from composting facilities in Germany. Cultured isolates of *A. fumigatus* produced a range of mycotoxins. Extracts of total dust and bioaerosols from a composting hall contained two *A. fumigatus* mycotoxins, two tryptoquivaline (which has tremorgenic properties) and trypacidin. They did not find the most toxic mycotoxins produced by *A. fumigatus* (gliotoxin and verruculogen) in the bioaerosols. In one study Fischer[45] did not find mycotoxins in samples of airborne spores, organic dusts and bioaerosols from composting sites. The results indicated that fungal spore counts need to be above 10^7 m^{-3} air to enable detection of fungal metabolites or toxins.

It is not known what occupational health risks are posed by the fungal metabolites in compost bioaerosols; their toxigenic potential needs further investigation.

Glucans

$(1\rightarrow3)$-β-D-glucan is a polyglucose compound in the cell walls of fungi, some bacteria and plants. It is a potent inflammatory agent that induces non-specific inflammatory reactions and may also be a respiratory immunomodulatory agent. Exposure to $(1\rightarrow3)$-β-D-glucans has been associated with an increased prevalence of atopy, decreases in forced expiratory volume (FEV$_1$), and adverse respiratory health effects in the indoor and occupational environment.[47–49] There is also evidence that $(1\rightarrow3)$-β-D-glucans may enhance pre-existing inflammation.[47]

$(1\rightarrow3)$-β-D-glucans may be involved in contributing to the inflammatory responses resulting in respiratory symptoms and adverse lung function effects in response to the inhalation of bioaerosols.

Volatile Organic Compounds

Volatile organic compounds (VOCs) are generated by many sources in the compost mixture including micro-organisms. Eitzer[50] found that most VOCs were emitted during the early stages of processing. Emissions were concentrated at the tipping floors where waste arrives, at the shredder and at the initial active composting region. VOC concentrations, even 'worst case' samples collected right next to the compost, were well below USA permissible workplace levels. They also found great similarities between facilities operating under differing

[45] G. Fischer, R. Ostrowski and W. Dott, *Chemosphere*, 1999, **38**, 1745–1755.
[46] G. Fischer, R. Schwalbe, M. Moller, R. Ostrowski and W. Dott, *Chemosphere*, 1999, **39**, 795–810.
[47] J. Douwes, I. Wouters and H. Dubbeld, *Am. J. Ind. Med.*, 2000, **37**, 459–468.
[48] P.S. Thorn and R. Rylander, *Am. J. Resp. Crit. Care Med.*, 1998, **157**, 1798–1803.
[49] D. Fogelmark and R. Rylander, *Int. J. Exp. Pathol.*, 1994, **74**, 85–90.
[50] B. Eitzer, *Environ. Sci. Technol.*, 1995, **29**, 896–902.

conditions, their sites ranging from aerated in-vessel systems to open windrows.

Fischer[46] screened 13 fungal species frequently isolated from a composting facility for microbial VOCs. They identified various hydrocarbons and terpenes. However, these are not necessarily produced when the fungi are growing in the environment.

A wide variety of VOCs can also originate from plant material. There is not enough information available on exposure to microbial VOCs to enable any assessment of potential health risks.

3 Potential Ill Health Effects among Compost Workers

The effects of occupational exposure to organic dusts on respiratory health have been investigated, but the mechanisms through which these respiratory effects are caused are not yet well understood.

Many of the micro-organisms found in dust generated during composting are known respiratory sensitisers. Fungi such as *Aspergillus* spp., *Penicillium* spp., *Cladosporium* spp., *Rhizopus* spp. and *Alternaria* spp. are well known allergens[28,51–53] while Gram-negative bacteria may also be a source of endotoxin.[54]

Inhalation of organic dust can cause a range of immunological respiratory symptoms which can be divided into four types of respiratory reaction as well as, very infrequently, infection.[29,55–57]

Allergic Rhinitis and Asthma

When a patient is sensitised to airborne allergens, exposure to those allergens can trigger the immunoglobulin E (IgE) pathway of the immune system causing allergic rhinitis (inflammation of the nasal passageways) or allergic asthma (upper respiratory tract broncho-constriction). Rhinitis and asthma frequently coexist in the same patient and both diseases are increasing in prevalence in the general population. Organic dust rhinitis and asthma are not caused by a single allergen present in the dust; different allergens may be responsible in different patients.[29,58–60] Workers handling compost are often exposed to higher levels of allergens than the general population and the species to which they are exposed may differ.

[51] J. Dutkiewicz, L. Kus, E. Dutkiewicz and C. P. W. Warren, *Ann. Allergy*, 1985, **54**, 65–68.
[52] J. Dutkiewicz, S. A. Olenchock, W. G. Sorenson, V. F. Gerencser, J. J. May, D. S. Pratt and V. A. Robinson, *Appl. Environ. Microbiol.*, 1989, **55**, 1093–1099.
[53] J. Lacey, *Ann. Agric. Environ. Med.*, 1995, **2**, 31–35.
[54] J. Dutkiewicz, *Zbl. Bak. I. Abt. Orig.*, 1976, **236**, 487–508.
[55] M. Chan-Yeung, D. A. Enarson and S. M. Kennedy, *Am. Rev. Respir. Dis.*, 1992, **145**, 476–487.
[56] J. Lacey, *Postharvest News Info.*, 1990, **1**, 113–117.
[57] R. Rylander, *Ann. Agric. Environ. Med.*, 1994, **1**, 7–10.
[58] B. Crook, *Grana*, 1994, **33**, 81–84.
[59] A. D. Blainey, M. D. Topping, S. Ollier and R. J. Davies, *J. Allergy Clin. Immunol.*, 1989, **84**, 296–303.
[60] E. Zuskin, E. N. Schachter, B. Kanceljak, J. Mustajbegovic and T. Wiltek, *Int. Arch. Occup. Environ. Health*, 1994, **66**, 317–324.

Chronic Bronchitis and Chronic Obstructive Pulmonary Disease

Chronic bronchitis is an inflammation of the mucous membrane of the bronchial tubes characterised by chronic cough, hypersecretion of phlegm and/or sputum and dyspnea and/or airways obstruction. The role of airborne fungal spores is uncertain, but airborne bacterial endotoxins may be involved in these diseases.[29,61,62] Clapp *et al.*[61] found evidence of additional endotoxin independent mechanisms of lung inflammation.

Extrinsic Allergic Alveolitis or Granulomatous Pneumonitis

Extrinsic allergic alveolitis or granulomatous pneumonitis is generally an occupationally-related disease. Extrinsic allergic alveolitis is a T lymphocyte CD8 predominantly, granulomatous, inflammatory reaction of the peripheral gas exchange tissue. Onset can be acute or insidious.[63] Repeated exposure to large concentrations of spores, in excess of 10^6 spores m^{-3} of air (and mostly 1–5 μm in diameter) has been suggested as the cause of acute symptoms.[55] However prolonged exposure to lower concentrations of spores may also cause chronic symptoms.[64] Acute symptoms occur 4–6 hours after exposure to the dust. These are chills, fever, dry cough, malaise and increasing breathlessness and, eventually, permanent lung damage may occur.[28,29,65] The characteristic immunological feature is the occurrence of predominantly IgG antibodies against specific antigens in the organic dust. Mushroom workers' lung has been linked to the inhalation of actinomycete spores associated with the mushroom compost.[66,67] Farmers' lung disease and is caused by inhalation of the fungal and actinomycete spores present in large numbers in contaminated organic dusts; the actinomycetes *Saccharopolyspora* (*Faenia*) *rectivirgula* and *Thermoactinomycetes* spp. have been implicated.[56]

Toxic Pneumonitis or Organic Dust Toxic Syndrome

Toxic pneumonitis or organic dust toxic syndrome (ODTS) is an acute illness occurring during, or shortly after, high exposure to airborne dust. Influenza-type symptoms develop, with leucocytosis and fever. No prior sensitisation is needed, antibodies do not develop and respiratory symptoms may or may not occur. The aetiology is unknown, but this also may be caused by inhalation of mycotoxins or endotoxins present in organic dust.[29,55] Further evidence for the role of endotoxins in ODTS has been provided by human challenge studies. Challenge

[61] W. D. Clapp, S. Becker, J. Quay, J. L. Watt, P. S. Thorne, K. L. Frees, X. Zhang, H. S. Koren, C. R. Lux and D. A. Schwartz, *Am. J. Resp. Crit. Care Med.*, 1994, **150**, 611–617.

[62] S. A. Olenchock, D. C. Christiani, J. C. Mull, T. T. Ye and P. L. Lu, *Biomed. Environ. Sci.*, 1990, **3**, 443–451.

[63] J. N. Fink, *J. Allergy Clin. Immunol.*, 1973, **52**, 309–317.

[64] J. Lacey and J. Dutkiewicz, *J. Aerosol. Sci.*, 1994, **25**, 1371–1404.

[65] S. Webber, G. Kullman, E. Petsonk, W. G. Jones, S. Olenchock, W. Sorenson, J. Parker, R. Marcelo-Baciu, D. Frazer and V. Castranova, *Am. J. Ind. Med.*, 1993, **24**, 365–374.

[66] J. G. Kleyn, W. M. Johnson and T. F. Wetzler, *Appl. Environ. Microbiol.*, 1981, **41**, 1454–1460.

[67] J. Lacey, P. A. M. Williamson and B. Crook, in *Aetiology*, ed. M. Muilenberg and H. Burge, CRC/Lewis Publishers, New York, 1996.

studies with pure endotoxin have shown that in normal healthy subjects an inhalation of over 30–300 μg endotoxin can cause a clinical response.[68–70] Inhalation of endotoxin can also result in decreases in lung function and inflammatory responses. A decrease in lung function is caused by inhalation of over 80 μg of endotoxin in healthy subjects and over 20 μg endotoxin in asthmatics;[70–72] this response is significant 30 minutes after endotoxin inhalation and lasts five hours or more. Inflammatory responses to an acute inhalation of LPS in healthy subjects have been reported to occur after inhalation of less than 0.5 μg. Endotoxins have strong adjuvant effects on the reactions to antigens and increase the production of antibodies, they can have a synergistic effect on the skin prick test response and may facilitate the development and persistence of hypersensitivity pneumonitis and allergic asthma *via* their inflammatogenic and adjuvant properties.[49,69,73] The adjuvant effect could be very important in the context of occupational exposure to the mixed allergens and toxins in compost dust.

4 Compost Site Case Studies

Wheeler *et al.*[74] investigated microbial emissions and worker health at three composting sites in the UK. These included one open windrow site processing green waste, an open windrow site processing mixed green and source separated household organic waste, and an in-vessel system processing mixed green waste, source separated household organic waste and refuse derived fuel production fines. In the investigation a range of aerobiological samplers were used to monitor airborne viable micro-organism levels during the different composting processes on different days. Handling of green waste compost in the open generated levels of airborne bacteria which exceeded 10^6 colony forming units (cfu; a measure of culturable microbial cells) m^{-3} air sampled on occasions. Levels of Gram-negative bacteria, fungi and actinomycetes each at times exceeded 10^5 cfu m^{-3} air sampled. The handling of mixed waste compost generated levels of airborne bacteria at times in excess of 10^5 cfu m^{-3}.

Gram-negative bacteria sometimes exceeded 10^5 cfu m^{-3}, fungi 10^4 cfu m^{-3} and actinomycetes 10^5 cfu m^{-3} air sampled. Levels of airborne bacteria were highest during shredding and turning, airborne fungi during screening and airborne actinomycetes during screening and shredding at the open sites. The in-vessel composting process generated levels of airborne bacteria and Gram-negative bacteria which both exceeded 10^7 cfu m^{-3} air. Fungi and actinomycetes both

[68] R. Rylander, B. Bake, J. J. Fischer and I. M. Helander, *Am. Rev. Resp. Dis.*, 1989, **140**, 981–986.

[69] R. Rylander, *Int. J. Occup. Environ. Health*, 1997, **3**, S32–S36.

[70] O. Michel, R. Gianni, B. Le Bon, J. Content, J. Duchateau and R. Sergysels, *Am. J. Resp. Crit. Care Med.*, 1997, **145**, 1157–1164.

[71] O. Michel, J. Duchateau and R. Sergysels, *Am. Physiol. Soc.*, 1989, 1059–1064.

[72] O. Michel, R. Gianni, B. Le Bon, J. Content, J. Duchateau and R. Sergysels, *Am. Rev. Resp. Dis.*, 1992, **146**, 352–357.

[73] O. Michel, R. Gianni, J. Duchateau, F. Vertongen, B. Le Bon and R. Sergysels, *Clin. Exp. Allerg.*, 1991, **21**, 441–448.

[74] P. A. Wheeler, I. Stewart, P. Dumitrean and B. Donovan, R & D Technical Report P1-315/TR, Environment Agency, Bristol, UK, 2001.

exceeded 10^4 cfu m^{-3}. Levels of airborne micro-organisms were highest during unloading of the vessel and at the biofilter at the invessel site.

Inhalable dust was generally low and reached 'normal' levels within 250 m of the site. VOCs were also generally low and well below the safety guidelines.

Occupational health monitoring was also carried out on 11 workers. These included skin checks, respiratory function tests, spirometry, blood counts, kidney and liver function assessment and urine analysis. The workers at the outdoor sites had no problems that might be associated with exposure to microbial emissions from the compost. However, the workers at the in-vessel site had adverse reactions to particular operations. For example, where they had to work within the composting vessel and were potentially exposed to very high levels of airborne dust and micro-organisms, symptoms were characteristic of ODTS.

Green Waste Sites

Lacey[75] reported the preliminary results from a UK study of actinomycetes in green waste compost. During the shredding of fresh green waste, concentrations of airborne actinomycetes were less than 5×10^4 cfu m^{-3}, and levels averaged 10^6 cfu m^{-3} close to the compost piles during turning. They found that *Saccharomonospora* was the predominant taxon during shredding, *Steptomyces albus* and *Thermoactinomyces* were also present. During composting other taxa such as *Steptomyces* increased in numbers.

Folmsbee[76] investigated levels of airborne micro-organisms at an outdoor composting centre in Oklahoma, USA. A sampling site downwind of the main composting activity showed a ten-fold increase in all the micro-organisms monitored in comparison with the other sites. The average concentration of total viable bacteria was 5059 cfu m^{-3}, Gram-negative bacteria were 2023 cfu m^{-3}, fungi were 972 cfu m^{-3}, and actinomycetes were 2159 cfu m^{-3}. Overall, fungal levels were lower than those of bacteria.

Heida *et al.*[77] investigated bioaerosols in an indoor green waste composting facility. Total bacteria counts were more than 28 000 cfu m^{-3}, Gram-negative bacteria were up to 9100 cfu m^{-3} and included the genera *Pseudomonas*, *Acinetobacter*, *Shigella* and *Yersinia*. Total fungal counts were less than 9000 cfu m^{-3} and included *Aspergillus* (60%), *Penicillium* (20%), *Mucor* and *Rhizopus*.

Hryhorczuk *et al.*[78] measured bioaerosol emissions from a green waste composting facility in Chicago. Concentrations of airborne bacteria, fungi, endotoxin and β-glucans were significantly higher on-site than off-site. Levels of bacteria 1 m downwind from the compost piles reached 7.9×10^4 cfu m^{-3} and levels of fungi reached 1.8×10^4 cfu m^{-3}. Endotoxin levels on site reached 6.06 ng m^{-3} (60 EU m^{-3}) and β-glucans reached 14.45 ng m^{-3}. The most predominant fungi were *Aspergillus* (45%) and *Penicillium* (21%); *Cladosporium* (8%) and *Alternaria* (5%) were also present. The composting facility activities significantly

[75] J. Lacey, *Ann. Agric. Environ. Med.*, 1997, **4**, 113–121.

[76] M. Folmsbee, *Air Waste Manage. Assoc.*, 1999, **49**, 554–561.

[77] H. Heida and S. Van der Zee, *Am. Ind. Hyg. Assoc.*, 1995, **56**, 39–43.

[78] D. Hryhorczuk, P. Scheff, J. Chung, M. Rizzo, C. Lewis, N. Keys and M. Moomey, *Ann. Agric. Environ. Med.*, 2001, **8**, 177–185.

increased downwind concentrations of bacteria off-site. While bacterial concentrations averaged 19 044 cfu m^{-3} one metre away from the compost piles on-site, the average was down to 5915 cfu m^{-3} at the boundary fence 75 m away from the nearest piles. The highest concentrations of total particulates, endotoxin, and β-glucans were observed in the personal samplers worn by the workers or attached next to the worker in an open-cab front end loader. These data were not reported separately and were included under on-site samples. The study also found that concentrations of total viable fungi and total fungal spores in the community adjacent to the composting facility were similar to outdoor fungal concentrations in other control communities. This study also found that the mean total viable fungi was higher off-site than on-site (8651 *vs.* 3068 cfu m^{-3}), the area surrounding the facility being wooded with wetland areas.

Municipal Biowaste Sites

Nielsen *et al.*[79] investigated micro-organisms and endotoxin in experimentally generated bioaerosols from composting source separated household waste. Their compost was made of 86% biowaste, 10% straw, 4% paper. Samples of 1, 5 and 9 week old compost were collected and used to generate bioaerosols. The micro-organisms in the compost were predominantly bacteria and actinomycete spores (total of 10^9–10^{11} cells g^{-1}). There was a significant increase in numbers between week 0 and week 9, and numbers increased steadily over the 11 week period. The fungal concentrations were low, yet *Aspergillus fumigatus* was the predominant viable fungal species. Fungi increased significantly at the end of the composting period but numbers were still just above the detection level (200 cfu g^{-1}) in most samples. Endotoxin levels varied, with a significant increase between the minimum measured (2.5 μg g^{-1}) at week 5 and the maximum (110 μg g^{-1}) at weeks 9–11. Actinomycete spores were found to be particularly prone to becoming airborne and the authors concluded that thermophilic actinomycetes were the predominant source of airborne micro-organisms.

Lavoie[80] investigated bioaerosols at two household waste sorting and composting plants in Canada, both sites comprising a reception area and a fermentation building. Site A had fermentation cylinders and indoor windrows, while site B had indoor windrows and outdoor curing. At both sites peak levels of airborne micro-organisms occurred in the reception and fermentation areas and this was regardless of season. At site A total bacteria reached 10^5 cfu m^{-3}, of which 10^2 cfu m^{-3} were Gram-negative bacteria. Actinomycetes reached 10^3 cfu m^{-3} and fungi 10^4 cfu m^{-3}, with peak concentrations of *A. fumigatus* also measured at 10^4 cfu m^{-3} of air sampled. At site B total bacteria reached 10^4 cfu m^{-3} total bacteria, of which 10^2 cfu m^{-3} were Gram-negative bacteria. Actinomycetes and fungi both peaked at 10^4 cfu m^{-3}, with peak concentrations of *A. fumigatus* at 10^3 cfu m^{-3} of air sampled. Levels of bacteria were over 10^4 cfu m^{-3} at several of the indoor sampling areas. In several areas total fungal concentrations were significantly higher than in the outdoor air, and the *A. fumigatus* levels measured in all workstations at both sites were significantly

[79] B. Nielsen, N. Breum and O. Poulsen, *Ann. Agric. Environ. Med.*, 1997, **4**, 159–168.
[80] J. Lavoie, *Ann. Agric. Environ. Med.*, 1997, **4**, 123–128.

higher than in the outdoor air nearby. Bioaerosol concentrations in air 100 m downwind from the sites were not affected by operations.

Tovalen *et al.*[81] investigated bioaerosols from composting source separated biowastes in Finland. The compost was processed outdoors. Concentrations of airborne microbes and endotoxin were highest during crushing of fresh waste and turning of compost. Both bacterial and fungal levels ranged between 10^3–10^5 cfu m^{-3}; levels were higher in the summer when the compost was dry. The predominant fungi were *Aspergillus* spp. and *Penicillium* spp. Concentrations of actinomycetes were lower than fungi during all stages of the composting process, 0–3000 cfu m^{-3}, compared to background concentrations less than 120 cfu m^{-3}. Endotoxin levels were high at 0.8–344.5 ng m^{-3} (about 80–3445 EU m^{-3}) outside and 0–152 ng m^{-3} (about 0–1520 EU m^{-3}) inside vehicle cabs.

The numbers of faecal streptococci, faecal coliform and *Clostridia* at different stages of the composting process were also investigated.[81] Levels were highest in fresh biowaste, 10^4–10^6 cfu g^{-1}, and, in well managed compost, decreased over the first 4 weeks to 150 cfu g^{-1}. In poorly managed compost the numbers increased over 6 weeks. In mature compost numbers of faecal streptococci were 100–300 cfu g^{-1}, faecal coliform 600–3800 cfu g^{-1} and *Clostridia* 4000–5000 cfu g^{-1}. Pathogenic bacteria such as *Escherichia coli*, *Salmonella typhimurium* and *Klebsiella* were found in the 2–4 week old compost. Psychrotrophic bacteria predominating at 2 months were overtaken by mesophilic bacteria at 3 months. Thermophilic bacteria numbers were high all the time and were most abundant in the fresh compost. As a result of these potential levels of exposure, they advised cleaning out the cabins of the machines at least twice a week.

Deportes *et al.*[82] also investigated the survival of pathogens in compost. They reported that at least one hour at 68 °C will kill most pathogens; for example *Salmonella* species require 15 to 20 minutes at 60 °C and *E. coli* and *Shigella* species require one hour at 55 °C. Other authors recommended that temperatures of 55–60 °C for at least three days will destroy most pathogens.[83,84] Otten *et al.*[85] detected no *E. coli* or *Salmonella* species in compost from source separated kitchen and yard waste.

Marchand *et al.*[86] undertook sampling at a municipal solid waste recycling and composting plant in Quebec, Canada. In at least one sample from each workstation, the air concentrations of total bacteria exceeded 10^4 cfu m^{-3} (maximum 5.3×10^5 cfu m^{-3}). The concentrations of Gram-negative bacteria were above 1000 cfu m^{-3} in the air from six out of nine workstations (maximum 7900 cfu m^{-3}) although the compost method used appeared to kill Gram-negative bacteria. The indoor concentrations of fungi were higher than outdoor concentrations in most areas (maximum 7200 cfu m^{-3}).

[81] O. Tovalen, A. Veijanen and K. Villberg, *Waste Manage. Res.*, 1988, **16**, 525–540.

[82] I. Deportes and D. Zmirou, *Sci. Total Environ.*, 1995, **172**, 197–222.

[83] E. Epstein and J. I. Epstein, *Biocycle*, 1989, **30**, 50–53.

[84] M. S. Finstein and F. C. Miller, *Composting of Agricultural and Other Wastes*, Elsevier Applied Science, Barking, UK, 1985, pp. 13–26.

[85] L. Otten, R. Voroney, J. Winter, S. Andrews, H. Lee and J. Trevors, *Health Effects of Composting*, AEA Technology, Oxfordshire, UK, 1999, 1–109.

[86] G. Marchand and L. Lazure, *Air Waste Manage. Assoc.*, 1995, **45**, 778–781.

In-vessel Composting

Danneberg *et al.*[87] examined microbial and endotoxin emissions in the neighbourhood of a Herhof box-system composting plant. They used a dispersion model and found no significant increase in microbial concentrations, compared to concentrations described in the literature for ambient air, at a distance of 500 m from the site.

Mushroom Compost

During the commercial production of mushroom compost, the highest levels of airborne micro-organisms are associated with handling compost, such as when the mushroom spawn is being mixed with the compost, or when disposing of spent compost after mushroom growth has been completed. The predominant micro-organisms are thermophilic actinomycetes. Aerobiological studies at mushroom farms have shown that during the mixing of compost with spawn thermophilic actinomycetes may be present in the air in enclosed facilities at concentrations from 10^7 to 10^9 cfu m^{-3}, with fungal concentrations from 10^3 to 10^5 cfu m^{-3}.[88,89] During the emptying of trays of spent compost, actinomycete concentrations were measured at 10^4 cfu m^{-3} and fungi at 10^5 cfu m^{-3},[88] while in an earlier study[66] total spore counts were made, estimating the airborne spore levels to be in the region of 10^5 m^{-3}, with actinomycetes predominating. These authors also estimated that a worker dumping spent compost could inhale 6×10^7 micro-organisms during the 2.5 hours taken to complete the task. Air samples taken out of doors in a mushroom compost preparation area revealed considerably smaller bioaerosol concentrations.[88] Between windrows of Phase 1 composting material numbers ranged from 10^3 to 10^5 cfu m^{-3}, compared to 10^2–10^3 cfu m^{-3} upwind of the site, showing the influence of the proximity of the composting material on the airspora.

Sewage Compost

Millner *et al.*[90] reported levels of 1.5×10^4 cfu m^{-3} thermophilic actinomycetes downwind of the turning of compost windrows containing a mixture of wood chips and sewage sludge. Actinomycetes identified included *Nocardia*, *Saccharopolyspora*, *Saccharomonospora* spp. and *Streptomycete* spp.

Very high airborne endotoxin levels were measured at an indoor municipal sewage sludge composting plant in Colorado.[91] Endotoxin levels ranged from 28.9 to 5930.6 ng m^{-3} (about 289–59 306 EU m^{-3}) which is 30 times greater than the threshold at which ODTS is thought likely to occur.[70] The maximum total dust level measured was 173.8 mg m^{-3}. Dust was generated by traffic, mixing the

[87] G. Danneberg, E. Grüneklee, M. Seitz, J. Hartung and A. J. Driesel, *Ann. Agric. Environ. Med.*, 1997, **4**, 169–173.
[88] B. Crook and J. Lacey, *Grana*, 1991, **30**, 446–449.
[89] H. G. G. van den Bogart, G. van den Ende, P. C. C. van Loon and L. J. L. D. van Griensven, *Mycopathologia*, 1993, **122**, 21–28.
[90] P. D. Millner, D. A. Bassett and P. B. Marsh, *Appl. Environ. Microbiol.*, 1980, **39**, 1000–1009.
[91] A. Darragh, D. Sandfort and R. Coleman, *Appl. Occup. Environ. Hyg.*, 1997, **12**, 190–194.

sludge, and screening operations at the composting facilities. The dust was composed of large cellulose materials and small sewage particles. A very large percentage of the aerosol was of a sufficiently small particle size to allow penetration into the lungs to the gas exchange region, where it can potentially cause the most damage.

Reducing the Level of Exposure to Bioaerosols

It may be possible to reduce emissions of bioaerosols, and therefore workers' exposure, by changing work practices. Fischer *et al.*[92] investigated the effect of turning frequency on levels of *Aspergillus fumigatus* during windrow composting of garden and kitchen waste. *A. fumigatus* levels in the centre of windrows were reduced after two weeks of composting from $> 10^3$ cfu g^{-1} dry weight compost (gdw) down to 10^2 cfu/gdw. Surface *A. fumigatus* counts remained high in less frequently turned windrows. Airborne concentrations of *A. fumigatus* were highest (6×10^5 cfu m^{-3}) while compost was being turned after eight weeks, but release of fungal spores was always 1–2 orders of magnitude lower if the compost had been turned daily or weekly during that time. The more frequently the windrow was turned (daily or weekly) the faster the temperature increased to a level which can eliminate *A. fumigatus*. They concluded that health risks for personnel working at composting sites could be lowered by frequently turning the windrows and thus reducing the levels of *A. fumigatus* on the surface of the compost. Investigation of concentrations 10 m downwind of turning process showed numbers already two to three orders of magnitude lower. Outside the perimeter of the site *A.* fumigatus levels were normally not higher than in control samples.

Epstein *et al.*[93] investigated an enclosed biosolids composting facility. When dust control measures (which included increasing the moisture content of the feed mix, altering air flow and putting in place misters), concentrations of dust, endotoxin and *A. fumigatus* were reduced by around 90%.

Breum *et al.*[94] used a dustiness drum to look at airborne micro-organisms and endotoxin of stored waste and showed that materials stored with free access to air gave significantly higher concentrations of airborne micro-organisms and endotoxin than those stored with low access. Concentrations of bioaerosols were correlated to the weight loss of water in storage.

Ambient Bioaerosol Levels and Dispersal from Compost Sites

'Normal' levels of micro-organisms.   The case studies described above show the levels of bioaerosols to which workers handling compost, and the people in the vicinity, may be exposed. However, to place these data in context, it is important to recognise that bioaerosols are constantly present in the ambient atmosphere as a consequence of dust from soil and the natural breakdown of vegetation. In

[92] J. Lott Fischer, T. Beffa, P.-F. Lyon and M. Aragno, *Waste Manage. Res.*, 1998, **16**(4), 320–329.
[93] E. Epstein, C. Youngberg and G. Croteau, *Compost Sci. Util.*, 2001, **9**, 250–255.
[94] N. O. Breum, B. H. Nielsen, E. M. Nielsen, U. Midtgaard and O. M. Poulsen, *Waste Manag. Res.*, 1997, **15**, 169–187.

addition, other human activities, such as various agricultural practices, will create bioaerosols.

Soil is a major source of micro-organisms and one of the most complex of natural communities. It is generally accepted that only a small percentage, possibly as few as 1%, of the micro-organisms present in soil are culturable, and there may be as many as 4000 species per gram of soil.[95] The numbers present will depend on the type of soil, organic loading, temperature and moisture content *etc.* However, as a general guide, bacterial density in soils, sludges and sediments can be considered to be in the range of 10^8–10^{10} gram^{-1}.[96] As a consequence, any dusts made airborne by the disturbance of soil will support large numbers of micro-organisms.

In the absence of any significant bioaerosol sources, natural atmospheric conditions in a typical suburban area were reported to give rise to 0–7.2×10^3 (mean 273) cfu m^{-3} mesophilic fungi, 0–193 (mean 2.1) cfu m^{-3} thermophilic fungi, 0–71 (mean 1) cfu m^{-3} *A. fumigatus*, 42–1.6×10^3 (mean 79) cfu m^{-3} bacteria. The highest concentrations occurred during summer and autumn.[97] Crook and Lacey[98] reported concentrations of viable airborne micro-organisms outdoors to be: 500 cfu m^{-3} total bacteria, 10 cfu m^{-3} Gram-negative bacteria, 1200 cfu m^{-3} total mesophilic fungi, 300 cfu m^{-3} thermophilic fungi and 60 cfu m^{-3} thermophilic bacteria and actinomycetes. Ambient levels of viable airborne bacteria in an agricultural area were reported by Bovallius *et al.*[99] as 2–3.4×10^3 (mean 99) cfu m^{-3}, and in a city 100–4.0×10^3 (mean 850) cfu m^{-3}. These levels are about 10–10^6 times lower than those recorded during the handling of compost. High levels of urban micro-organisms have been recorded downwind of a wetland and wooded area.[78]

Dispersal. Bioaerosols are formed during the composting process whenever materials are agitated mechanically. However, concentrations have been shown to decrease to background after site activities cease, suggesting that windblown aerosolisation is insignificant.[90,100] As bioaerosol masses are typically small (hence they have small settling velocities), they can be carried long distances by the wind and thermal currents.[101] The pattern of dispersal of bioaerosols around a composting site depends upon a number of factors, including the emission rate (the number of micro-organisms liberated per unit time), prevailing atmospheric conditions (wind speed and direction, solar incidence, temperature gradients and relative humidity[102,103] and local topography which will determine the air flow

[95] A. Ogram and X. Feng, in *Manual of Environmental Microbiology*, ed. C. J. Hurst, G. R. Knudsen, M. J. McInerney, L. D. Stetzenbach and M. V. Walter, American Society for Microbiology Press, Washington, pp. 422–430.

[96] W. E. Holben, in *Manual of Environmental Microbiology*, ed. C. J. Hurst, G. R. Knudsen, M. J. McInerney, L. D. Stetzenbach and M. V. Walter, American Society of Microbiology Press, Washington, pp. 431–436.

[97] B. L. Jones and J. T. Cookson, *Appl. Environ. Microbiol.*, 1983, **45**, 919–934.

[98] B. Crook and J. Lacey, *Environ. Technol. Lett.*, 1988, **9**, 515–520.

[99] A. Bovallius, B. Brucht, R. Roffey and A. Anas, *Appl. Environ. Microbiol.*, 1978, **35**, 847–852.

[100] F. J. Passman, *Mycopathologia*, 1983, **83**, 41–51.

[101] H. A. McCartney, in *Species Dispersal in Agricultural Habitats*, ed. R. G. H. Bunce and D. C. Howard, Bellhaven Press, London, 1990.

around the site.[104] The emission rate will depend on the process carried out, the type of machinery used, the moisture content of the compost, the microbial content of the material processed, and whether or not the process is enclosed or carried out in the open air.

Concerns have been expressed by residents neighbouring composting facilities, citing potential adverse health effects resulting from inhalation of bioaerosols from the site.[105,106] A number of studies have set out to assess the impacts composting facilities may have; however, monitoring bioaerosols in an outdoor environment is beset with practical difficulties.[107] Consequently, there have been few detailed studies reported in the scientific literature that relate specifically to composting. By contrast, there are numerous reports on the dispersal of pollen and spores, especially those that are of agronomic importance (see, for example, McCartney),[108] which are thought to decrease in concentration away from the source term following either the power law or exponential models.[109]

Millner *et al.*[90] and Lighthart and Mohr[102] both suggested that wind turbulence played a key role in the dispersal of airborne micro-organisms. Millner *et al.* monitored concentrations of *A. fumigatus* at a sewage sludge composting facility and applied a Pasquill dispersion model to the data. It predicted that background concentrations of *A. fumigatus* would be attained 500–600 m from the source under unstable (turbulent) conditions, with distances in excess of 1 km under stable conditions. Lighthart and Mohr used a simulated virus (whose properties were derived from two actual viruses) and applied a Gaussian model. Their data suggested that turbulence dramatically affected downwind concentrations, with dilutions of 10^{-4} suggested within 30 m downwind from the source.

Similar decreases in concentrations of bioaerosols at composting facilities have been measured directly. Lacey and Williamson[110] observed a reduction of airborne fungi and bacteria to concentrations of less than 10% of those measured within 1 m of a turned pile of compost. Beffa *et al.*[111] noted a 100–1000-fold decrease in concentrations of *A. fumigatus* 10 m from a turning machine; at 500 m downwind, concentrations of 0–20 cfu m^{-3} were measured.

Passm an[100] monitored concentrations of *A. fumigatus* at composting facilities in Maine, USA. Samples taken 150 m downwind were not above background

[102] B. Lighthart and A. J. Mohr, *Appl. Environ. Microbiol.*, 1987, **53**, 1580–1583.

[103] B. Lighthart, *FEMS Microbiol. Ecol.*, 1997, **23**, 263–274.

[104] G. M. Heisler and D. R. Dewalle, *Agric. Ecosystems Environ.*, 1988, **22/23**, 41–69.

[105] P. D. Millner, S. A. Olenchock, E. Epstein, R. Rylander, J. Haines, J. Walker, B.L. Ooi, E. Horne and M. Maritato, *Compost Sci. Util.*, 1994, **2**, 6–57.

[106] M. L. Browne, C. L. Ju, G. M. Recer, L. R. Kallenbach, J. M. Melius and E. G. Horn, *Compost Sci. Util.*, 2001, **9**, 242–249.

[107] J. Lacey, P. A. M. Williamson and B. Crook, in *Aerobiology*, ed. M. Muilenberg and H. Burge, CRC Press, Boca Raton, Florida, 1996, pp. 1–17.

[108] H. A. McCartney, *Grana*, 1994, **33**, 76–80.

[109] B. D. L. Fitt, P. H. Gregory, A. D. Todd, H. A. McCartney and O. C. MacDonald, *J. Phytopathol.*, 1987, **117**, 227–242.

[110] J. Lacey and P. A. M. Williamson, CWM/110/93, Report to UK Department of the Environment Wastes Technical Division, 1995.

[111] T. Beffa, F. Staib, J. Lott Fischer, P.-F. Lyon, P. Gumowski, O. E. Marfenina, S. Dunoyer-Geindre, F. Georgen, R. Roch-Susuki, L. Gallaz and J.-P. Latgé, *Med. Mycol.*, 1998, **36**, 137–145.

concentrations, whilst at one site background concentrations were measured at 90 m downwind. Reinthaler *et al.*[112] measured concentrations of airborne micro-organisms at a number of waste handling and treatment facilities in Austria. At one composting site concentrations of bacteria were greater at 700 m than at 500 and 600 m, which was attributed to vehicle movements. Increased counts in residential areas adjoining the facility were ascribed to neighbouring farms. However, based on the data collected, the authors concluded that significantly lower counts were measured at distances greater than 200 m from the source.

Two hundred metres was the distance in which concentrations of *A. fumigatus* and total mesophilic bacteria were found to reach background concentrations by Gilbert *et al.*[113] This was used as the basis for the recommendation that routine sampling at a composting facility should be carried out if a 'sensitive receptor' lies within 200 m of the site boundary.[114]

In an attempt to assess potential adverse impacts, Millner *et al.*[105] conducted a comprehensive review of published data. The authors considered a number of different studies where downwind bioaerosol concentrations had been measured, and concluded: 'the data have indicated that at distances of 250–500 feet [76–152 m] from the compost facility perimeters the airborne concentrations of *A. fumigatus* were at or below background concentrations'.

By contrast the waste regulator in England and Wales (the Environment Agency) established a position in 2001 against 'permitting any new composting process (or modification to an existing process) where the boundary of the facility is within 250 metres of a workplace or the boundary of a dwelling, unless the application is accompanied by a site-specific risk assessment, based on clear, independent scientific evidence which shows that the bioaerosol levels are and can be maintained at appropriate levels at the dwelling or workplace'. The basis for this statement was work carried out by Wheeler *et al.*[74] who monitored bioaerosol dispersal from three composting facilities in England. Samples were collected downwind at the facilities using filters held in personal dust samplers designed for use in occupational exposure assessments. The data were modelled using the United States Environment Protection Agency SCREEN3 model, and estimates of the distances to assumed reference concentrations of 1000 cfu m^{-3} for total bacteria, 1000 cfu m^{-3} for total fungi, and 300 cfu m^{-3} for Gram-negative bacteria were made. Despite modelling positive slopes (increases in concentrations with distance from source) on a number of instances (which were ascribed to additional emissions and probably also reflected the small sample sizes), the authors concluded that concentrations generally reached the reference levels within 250 m of the source. Wheeler *et al.*[74] did, however, observe that many of the bioaerosols formed aggregates large enough to exhibit non-gaseous behaviour. It was suggested therefore that concentrations would decline with distance at a

---

[112] F. F. Reinthaler, D. Haas, G. Feierl, R. Schlacher, F. P. Pichler-Semmelrock, M. Köck, G. Wüst, O. Feenstra and E. Marth, *Zentralbl. Hyg. Umweltmed.*, 1998/99, **202**, 1–17.

[113] E. J. Gilbert, A. Kelsey, J. D. Karnon, J. R. M. Swan and B. Crook, *Proceedings of the 2002 International Symposium on Composting and Compost Utilization*, Ohio, USA.

[114] E. J. Gilbert and C. W. Ward, *Standard Protocol for the Sampling and Enumeration of Airborne Micro-organisms at Composting Facilities*, The Composting Association, Coventry, UK, 1999.

greater rate than a Gaussian dispersal model would predict.

Danneberg *et al.*[87] collected bioaerosol samples adjacent to a biofilter and trommel screen and at downwind sampling points at 150 m at a composting facility in Germany. They used the data to predict dispersal using a German TA Luft model, and concluded that typical background concentrations would be reached at 500 m.

There are a number of reasons why the use of mathematical models provides uncertain results when used to predict bioaerosol dispersal. Most have been developed to simulate atmospheric transport of particles over medium to long ranges (for example, estimating the release of particles from smoke stacks), with little information available on short range (< 1 km) dispersal. Determining the emission rate (source term) is complex, which stems from the fact that it is often moving and intermittent, and there are no direct methods available at present to measure it directly. Most are based on estimates of emission concentrations.[87,90] Additionally, buoyancy effects caused by the release of hot air into the cooler atmosphere have not yet been accounted for within any reported models. Coupled with the high degree of variability of naturally occurring background concentrations[97] and the effects of traffic movements and neighbouring bioaerosol sources, such as agricultural activities, the impacts composting facilities have on neighbouring populations requires further research.

Summary

Levels of airborne micro-organisms generated during the handling of compost can vary greatly from site to site depending on the scale and type of operation, and as a result different studies have reported widely differing levels. Most studies of micro-organisms associated with composting measure only the viable micro-organisms present. However, it is important to take into consideration that many of the ill health effects associated with inhalation of organic dusts can be caused by non-viable as well as viable micro-organisms. The studies reported here therefore underestimated the exposure to total (viable and non-viable) airborne micro-organisms. Microscopic counting has shown that plate counts of culturable micro-organisms can underestimate the numbers of workplace airborne micro-organisms by 10–20%.[115] Marchand *et al.*[86] found that the composting method they monitored appeared to kill the Gram-negative bacteria; their maximum counts of viable Gram-negative bacteria were well over 10^3 cfu m^{-3} but levels including the dead Gram-negative bacteria would have potentially been much higher. The predominant fungi measured at most composting sites are *Penicillium* spp. and *Aspergillus* spp.[77,78,86,98] Both are naturally present in the atmosphere but are in greater numbers near to actively turned compost.

It is very important that the composting process is properly managed. Levels of faecal coliforms, faecal streptococci and other potential pathogens in municipal waste have been shown to be greatly increased in badly managed compost.[81]

The majority of studies show that the bioaerosols concentrations generally returned to background levels 100–500 m from the site and many reach

[115] G. Blomquist, *Analyst*, 1994, **119**, 53–56.

background levels within 250 m from the compost.

Different sampling methods, different siting of samplers and other method variations all mean that results may differ from study to study. Even at one particular site concentrations can vary greatly from hour to hour by more than 10-fold.[116] Different composting activities have a dramatic effect on the levels of microbial emissions, as can weather conditions and wind speed and direction.[117] The moisture content of the compost also affects the bioaerosol levels.[74] This makes it difficult to predict 'typical' emissions of bioaerosols from composting operations and to equate this with respiratory hazard. However, an overall pattern is for a higher than normal exposure to bacteria, including Gram-negative bacteria, actinomycetes and fungi which have known potential to cause ill health.

5 Ill Health Case Studies

Few studies combine health data and exposure data, making any conclusions about dose responses hard to draw.

Sensitisation

Bunger *et al.*[118] carried out a cross sectional study, in Germany, to look at work related health complaints and immunological markers of exposure to bioaerosols among biowaste collectors and compost workers. 58 compost workers (mean duration of employment 3 years), 53 biowaste collectors (mean duration of employment 1.5 years) and 40 controls took part. The levels of specific IgG antibodies to fungi and bacteria were measured as immunological markers of exposure to bioaerosols. At the composting plants non-compostable materials were removed by manual sorting, the biowaste was mixed with shredded garden waste and piled in rows and the finished compost was sieved.

The levels of specific IgG antibodies to *A. fumigatus, A. nidulans, A. niger, A. versicolor, Penicillium* spp. *Saccharopoylyspora hirsuta, Saccharopolyspora rectivirgula, Saccharomonospora viridis* and *Streptomyces thermovulgaris* were measured.

The compost workers were found to have significantly more symptoms and diseases of the airways and skin than the control subjects. These included tracheobronchitis, mucous membrane irritation, sinusitis, eczema, dermatomycosis, pyoderma, nausea and ear inflammation. One compost worker complained of typical ODTS symptoms. Severe cases of infection or EAA or asthma were not found. Twenty compost workers had one or several increased antibody concentrations compared with only three biowaste collectors and one control. Significantly higher antibody titres to *A. fumigatus* were measured in workers at the composting plants. Compost workers also had higher titres to the other

[116] A. Neef, A. Albrecht, F. Tilkes, S. Harpel, C. Herr, K. Liebl, T. Eikmann and P. Kampfer, *Schriftenr. Ver. Wasser Boden Lufthyg.*, 1999, **104**, 655–664.

[117] R. Hofman, R. Bohm, G. Danneberg, S. Gerbi-Rieger, E. Gottlich, A. Koch, M. Kuhner, V. Kummer, K. Leibl, W. Martens, T. Missel, A. P. U. Neef, R. Rabe, B. Schilling, F. Tilkes and P. Wieser, *Schriftenr. Ver. Wasser Boden Lufthyg.*, 1999, **104**, 1–80.

[118] J. Bunger, M. Antlauf-Lammers and T. G. Schulz, *Occup. Environ. Med.*, 2000, **57**, 458–464.

fungal antigens compared to biowaste collectors and control subjects. Significantly increased antibody titres were also obtained for *Saccharopolyspora rectivirgula* and *Streptomyces thermovulgaris*. The concentrations for *Saccharopolyspora hirsuta* were also increased. There was a significant association between diagnosed diseases and increased IgG antibodies in the compost workers. There was also significant association between the duration of employment of the compost workers and the increased IgG titres, suggesting progressive development of IgG antibody responses with duration of exposure.

The workers involved in this study had been in the industry for a relatively short length of time (three years). Longitudinal studies into the long-term health effects of exposure to compost bioaerosols are needed to investigate whether workers who develop IgG against the allergens to which they are occupationally exposed go on to develop occupationally related symptoms.

Brown *et al.*[119] reported a case study of a man who developed hypersensitivity pneumonitis after working on the compost in his garden. Symptoms included respiratory difficulty about two hours after commencing work, fatigue, a non-productive cough, fever, chills, and pain in the joints. The patient made a full recovery within a few days, but the symptoms reoccurred on two subsequent occasions. The patient was found to have precipitating antibodies to *Thermoactinomyces vulgaris* and positive skin prick test to *Aspergillus*. The patient also had precipitating antibodies against an extract made from his compost pile. The patient had spent some considerable time gardening with the compost, working long hours every weekend. However, no information was reported on his possible exposure levels.

Upper Airway Inflammation

Douwes *et al.*[47,120] carried out a small study on 14 Dutch compost workers and 10 controls. For the duration of the study, nasal lavage was performed before and after the work shift on Mondays and Fridays. Mean personal dust and endotoxin exposures ranged from 0.4 to 3.1 mg m^{-3} and 50 to 100 EU m^{-3} respectively, and glucans ranged from 0.36 to 4.85 μg m^{-3}. Fungi levels were indicated to be over 10^6 cfu m^{-3}, total bacteria 10^9 cfu m^{-3} and Gram-negative bacteria 10^4 cfu m^{-3}. There was a cross shift increase in total cells and inflammatory mediator levels in the workers and a decrease in the controls. Total cells and inflammatory mediator levels in the workers were elevated pre-shift on Mondays compared to controls, and were more elevated at higher than lower endotoxin exposures. Occupational exposure in the compost workers was found to cause acute and (sub-)chronic non-immune or type III allergic inflammation in the upper airways. The authors suggest this is induced by non-allergic pro-inflammatory agents such as endotoxins and $\beta(1\rightarrow3)$-glucans.

[119] J. E. Brown, D. Masood, J. I. Couser and R. Patterson, *Ann. Allergy, Asthma Immunol.*, 1995, **74**, 44–47.
[120] J. Douwes, H. Dubbeld, L. van Zwieten, I. Wouters, G. Doekes and D. Heederik, *Ann. Agric. Environ. Med.*, 1997, **4**, 149–151.

Cytotoxic Effect of Compost

Non-immunological factors, such as cytotoxic effects, can also cause inflammatory responses. Cytotoxic responses have been linked to work related ill health in workers exposed to organic dusts. Roepstorff and Sigsgaard[121] used a cytotoxic assay to test a range of organic dusts and found that the most aggressive dusts tested were those with a high microbial content. Compost dust (5 week old organic household waste compost) and grain dust exerted an effect at very low concentrations after two hours incubation with monkey kidney cells and human lung carcinoma cells. The dusts tested, in order of decreasing cytotoxic effect, were compost, grain, swine and cotton. Further tests confirmed that the greatest cytotoxic potential occurred when the microbial activity was at its height in the composting process. However, pure endotoxin did not show any cytotoxic activity in the assay.[122] Further investigations are required to find out which microbially associated components are responsible for the cytotoxic potential.

General Ill Health

The outcome of a large study on the health of a population living near to a grass and leaf composting plant at Islip, New York was recently reported.[106] Sixty-three people living near the site and 82 controls were asked to keep a symptom diary. Individual personal exposure data were not collected, but bioaerosols were measured at fixed sites. Daily maximum *A. fumigatus* counts ranged from 30 to 19 000 spores m^{-3}. Average counts were 50 spores m^{-3} at the control neighbourhood, 100 spores m^{-3} in the study neighbourhood and 500 spores m^{-3} at the composting facility. Elevated spore counts (counts exceeding 300 spores m^{-3}) occurred in 15% of the counts in the study neighbourhood, in nearly 20% of the counts at the composting facility and in less than 5% of the counts at the control sites. When the study neighbourhood was downwind of the composting facility spore counts averaged four times the average background level. There was no evidence of *A. fumigatus* being associated with increases in respiratory or irritative symptoms, but there was an association with ragweed pollen (a common inhalant allergen), ozone, temperature, and time since the start of the study. Within the size limitation of the study, it was concluded that any major increase in allergy and asthma symptom prevalence in people living near the site was too small to detect, even though residents were exposed to elevated concentrations of *A. fumigatus* as a result of operations at the compost facility. Recer *et al.*,[123] reporting on the same study, recommend that since composting facilities are a permanent source of *A. fumigatus* spore emissions, this should be considered when siting open air composting sites in heavily developed areas.

Cobb *et al.*[124] investigated health complaints associated with commercial processing of mushroom compost. They could not demonstrate a significant health hazard. A comparison group, with no exposure to compost, had similar

[121] V. Roepstorff and T. Sigsgaard, *Ann. Agric. Environ. Med.*, 1997, **4**, 195–201.

[122] V. Roepstorff and T. Sigsgaard, *Waste Manage. Res.*, 1997, **15**, 189–196.

[123] G. Recer, E. Horn, K. Hill and W. Boehler, *Aerobiologia*, 2001, **17**, 99–108.

[124] N. Cobb, P. Sullivan and R. Etzel, *J. Agromed.*, 1995, **2**, 12–25.

symptoms to those living within 3000 feet [914 m] of the site.

Marth *et al.*[125] examined the occupational health of 137 employees at different waste handling facilities, including two composting facilities and three waste sorting plants. A medical examination, questionnaire and IgE measurements were carried out. No statistically significant increase of allergic diseases was found. There were no differences in lung function between workers and a control group. However, workers complained of hoarseness (38%), cough (35%), respiratory infections (23%), diarrhoea (18%), joint and muscle disorders (13%) and conjunctivitis (12%).

Infection

Aspergillosis. Under extreme circumstances, such as immunosuppression, *A. fumigatus*, an opportunistic pathogen, can cause infection.

Leenders *et al.*[126] found that an increase in numbers of patients with invasive aspergillosis could not be explained by an increase in the number of *Aspergillus* conidia in the outside air. They found that the outside air contained $0-9$ cfu m^{-3} with *A. fumigatus* numbers relatively constant, decreasing only from January to April.

Gastric infections. Ivens *et al.*[127] carried out a telephone questionnaire survey of 28 composting employees working at seven plants covering household and garden waste. 11% reported nausea and 11% reported diarrhoea, and there was a non-significant association between working with compost and diarrhoea. Among waste collectors the groups with highest exposure to total fungi or total micro-organisms reported fewer symptoms compared to the lower exposed groups. No positive trend was found, although there was an association between fungal exposure and diarrhoea.

Ivens *et al.*[128,129] further investigated the relationship between the gastrointestinal problems and bioaerosol exposure among waste collectors. An exposure–response relationship was found between nausea and endotoxin exposure and between diarrhoea and exposure to both endotoxins and viable fungi. Viable fungal spores reached levels greater than 10^7 cfu m^{-3}, total fungal spores (viable and non-viable) reached levels greater than 2×10^7 cells m^{-3} and endotoxin levels reached more than 500 EU m^{-3}. Bacteria levels were also very high, with total (viable and non-viable) micro-organisms exceeding 6×10^7 cells m^{-3}.

Sigsgaard *et al.*[38] investigated the health of eight compost workers. None had any skin problems and only one tested positive to a skin prick test against 10 common inhalant allergens. Mean total dust, total micro-organisms and endotoxin were 0.62 mg m^{-3}, 5.44×10^3 cfu m^{-3} and 0.8 ng m^{-3} respectively.

[125] E. Marth, F. F. Reinthaler, K. Schaffler, S. Jelovcan, S. Haselbacher, U. Eibel and B. Kleinhappl, *Ann. Agric. Env. Med.*, 1997, **4**, 143–148.

[126] A. Leenders, M. Behrendt, A. Luijendijk and H. Verbrough, *J. Clin. Microbiol.*, 1999, 1752–1757.

[127] U. Ivens, O. Poulsen and T. Skov, *Ann. Agric. Environ. Med.*, 1997, **4**, 153–157.

[128] U. Ivens, N. O. Breum, N. Ebbehoj, B. H. Nielsen, O. Poulsen and H. Wurtz, *Scand. J. Work Environ. Health*, 1999, **25**, 238–245.

[129] U. Ivens, J. Hansen, N. O. Breum, N. Ebbehoj, M. Nielsen, O. Poulsen, H. Wurtz and T. Skov, *Ann. Agric. Environ. Med.*, 1997, **4**, 63–68.

J. R. M. Swan, B. Crook and E. J. Gilbert

Gastrointestinal symptoms, and ever having experienced vomiting or diarrhoea in relation to work, were significantly more common in the composting industry than in controls.

Summary of Ill Health

There is a low incidence of serious chronic work related disease in compost workers (*i.e.* asthma, extrinsic allergic alveolitis). However, several of the studies show the 'early' responses to the microbial exposures of development of raised levels of IgG and inflammatory mediators in the workers.[118,130] Several have also reported a link between gastrointestinal symptoms and working with compost.[38,128] More information is required on whether compost workers who have developed raised levels of IgG and inflammatory mediators go on to develop work related diseases/ill health.

Although several of the studies reporting ill health among compost workers also recorded very high exposure to bioaerosols, few studies have looked at the personal exposure levels of the workers on the site. More personal monitoring is needed to establish peak exposure levels which could trigger allergic response and to provide some dose–response data for this industry.

There is not enough information available on the dose–response relationships between microbial exposure and development of ill-health. The differing susceptibility of individuals exposed also confuses the matter. Different people have differing susceptibility to microbially induced ill health. Some people, *e.g.* atopics, have a higher risk of becoming sensitised to allergens in the workplace even at 'low' concentrations.

It must be noted that in many of these studies, only viable micro-organisms were counted. Non-viable micro-organisms can also cause allergic reactions and this can be a complicating factor when trying to estimate dose–response effects. Eduard[131] estimated that IgG antibodies can be detected in response to 10^5 viable spores and 10^4 non-viable spores. The results from the airborne micro-organism surveys reported here show that the workers involved in composting are potentially regularly exposed to more than 10^5 viable spores.

Very high levels of airborne endotoxin were recorded at some sites.[82,91,128] Darragh et al.[91] recorded levels up to 59 306 EU m^{-3}. Some were well over the levels shown to cause acute ill health responses and with the potential to contribute to the development of chronic diseases.[69] At present there are no occupational exposure limits for endotoxin. The Dutch recommended health based occupational exposure limit for airborne endotoxin is 50 EU m^{-3} (4.5 ng m^{-3}) based on personal inhalable dust exposure measured as an eight hour time weighted average.[132] Rylander[69] recommends a no-effect level of 10 ng m^{-3} (100 EU) endotoxin. Microbial emissions from composting sites are a complex mixture and there is the potential of adjuvant effects when inhaling this mixture of toxins and allergens.

As the micro-organisms associated with compost are such common environ-

[130] J. Douwes and D. Heederik, *Int. J. Occup. Environ. Health*, 1997, **3**, S37–S41.
[131] W. Eduard, *Ann. Agric. Environ.*, 1997, **4**, 179–186.
[132] DECOS (Dutch Expert Committee on Occupational Standards), *Gezondheidsraad*, 1998.

mental contaminants that the general population are exposed to in low levels during their normal daily activities, it is very important that control populations – non-occupationally exposed populations, or populations sited some distance from composting activities – are included in studies aiming to collect dose–response data.

Containing the compost in buildings or vessels decreases emissions to the general population but potentially increases compost worker exposure. There is very little data publishe d on exposure of workers and the public due to premises that use in-vessel composting methods.

Few health based studies have recorded the duration of employment of the workers. Large-scale composting is a relatively new, rapidly expanding industry and some ill health effects may not yet have had time to develop. There is also some evidence of a healthy worker effect occurring. It may be that we are not yet seeing the health effects of long term exposure to low/medium levels of microbial emissions of this type.

The people working directly with the compost are exposed to significantly higher levels of bioaerosols but there is, as yet, very little evidence that this is causing ill health. If the workers who have a somewhat higher exposure to the microbial emissions generated by compost are not suffering from ill health effects then the general public who receive a much lower exposure are unlikely to suffer ill health effects. There is very little evidence that people living more than 250 m from composting sites are exposed to microbial emissions that are significantly higher than can be reached 'normally'. However, there are a few recordings of high levels. In one study, increased concentrations were detected 500 m away from a composting site, with concentrations of thermophilic actinomycetes reaching 10^6 cfu m^{-3} 200 m away.[116] Due to the *A. fumigatus* content of the emissions there may be an increased risk to immunocompromised persons sited downwind of a composting site.

Millner *et al.*,[105] reporting on the discussions at a workshop of experts, posed the question 'Do bioaerosols associated with the operation of biosolids or solid waste composting facilities endanger the health and welfare of the general public and the environment?'. They concluded that 'Composting facilities do not pose any unique endangerment to the health and welfare of the general public'. These conclusions were based on data showing that compost workers, who are exposed to the more concentrated bioaerosols, suffer from very few work related ill health effects. The airborne micro-organism level for most compost sites, whilst several fold higher than normal outdoor levels, are generally lower than in some occupational industries such as the grain industry where exposure to airborne micro-organisms can regularly be well over one million cfu m^{-3}.[133,134] However, any exposure to airborne micro-organisms must consititute some risk to respiratory health and there is a need to assess the risks posed by composting facilities, whether actual or perceived, in the context of risks presented by other environmental hazards. Clearly further monitoring and epidemiological data are required.

[133] J. R. M. Swan and B. Crook, *Ann. Agric. Environ. Med.*, 1998, **5**, 7–15.
[134] J. Dutkiewicz, *Eur. J. Resp. Dis.*, 1987, **71**, 71–88.

Health Effects and Landfill Sites

ANDY REDFEARN AND DAVE ROBERTS

1 Introduction

The potential for a causal link between landfill and certain adverse health outcomes in neighbouring residents is a matter of continuing concern. Establishing whether or not such a link actually exists is a complex process involving various areas of expertise, including public health, environmental science and waste management. The process encompasses epidemiological studies that examine incidence rates of adverse health outcomes in the vicinity of landfills and exposure risk assessment studies that examine whether predicted or measured exposures to harmful pollutant emissions from landfills are sufficient to cause adverse health outcomes.[1-3] Most studies to-date have been epidemiological in nature and cannot in isolation confirm a causal link, although epidemiological studies are often promoted by the press and pressure groups as showing absolute evidence.

This chapter reviews the evidence from both epidemiological and exposure risk assessments. The first section provides an overview of the epidemiological studies pertaining to landfill sites and identifies those health outcomes that have been rightly or wrongly most consistently linked to landfills. It is noted that many studies frequently cited in the context of landfills actually refer to miscellaneous waste sites that are not landfills at all or are unauthorised or unregulated waste dumps. This review pertains only to landfills defined as disposal sites designed and managed for the specific purpose of controlled disposal of waste.

The second section provides a more detailed review of recent epidemiological studies on adverse birth outcomes, including congenital anomalies (birth defects) and low birth weight, the health outcomes most consistently linked to landfills. The third section reviews the evidence from exposure risk assessment studies, focusing primarily on inhalation exposure to gaseous emissions because this appears to be the pathway of primary potential concern.

The premise of this review is that confirmation of the existence or magnitude of

[1] H. M. P. Fielder, C. M. Poon-King, S. R. Palmer, N. Moss and G. Coleman, *Br. Med. J.*, 2000, **320**, 19.
[2] H. Dolk, *Br. Med. J.*, 2000, **320**, 23.
[3] A. Redfearn, J. C. Dockerty, and R. D. Roberts, *IWM Scientific and Technical Review*, 2000, April, 14.

Issues in Environmental Science and Technology, No. 18
Environmental and Health Impact of Solid Waste Management Activities
© The Royal Society of Chemistry, 2002

a causal link between landfill and adverse health effects requires that the evidence satisfies the following criteria: (1) the epidemiological studies purporting adverse effects must be rigorously designed such that the health outcome data can be seen to provide convincing evidence, (2) there must be some degree of consistency amongst different epidemiological studies in terms of the types and significance of purported adverse health effects, (3) there must be a theoretical basis for the purported effects in terms of the likely mechanism or exposure pathway and (4) there must be a reality basis for the effects, as shown by actual field measurements in the ambient environment.

2 Overview of Epidemiological Studies

In this section the body of epidemiological literature on landfill sites is reviewed and the question is posed, taken as a body of research as a whole, do the studies on landfill sites indicate that there are specific health outcomes that consistently exhibit significant effects?

It is helpful to distinguish single-site and multi-site epidemiological studies. Single-site epidemiological studies attempt to determine whether there is an increased incidence of health effects in people living near individual landfill sites. This is done either by comparing disease incidence in the local population with incidence elsewhere (geographical studies); by comparing exposures of people who have disease symptoms with those who do not (case-control studies); or by following the disease incidence of exposed people through time (cohort studies). Single-site studies are prone to recall and reporting bias because they are often initiated in response to health concerns raised by the local community.[4] In addition, they are frequently limited in statistical power because of the small sample size available from the local population, particularly for rare health outcomes.

Multi-site studies typically select sets of waste sites and compare the frequencies of symptoms amongst people living near those sites with the frequencies amongst the general population. These studies attempt to resolve the difficulties of single-site studies by selecting large numbers of sites independently of community concerns or reported disease clusters.[4] Such studies have the additional advantage of having a large number of subjects, and therefore a greater ability to detect small increases in risk, *i.e.* increased statistical power. However, the use of such large-scale approaches has the disbenefit that detailed site-specific information on waste disposal histories, exposures and potential alternative pollution sources is seldom available.

There is an increasing body of epidemiological literature on the potential links between waste sites and health. In excess of 50 studies have been referenced since the mid-1980s. The studies have recently been the subject of comprehensive reviews elsewhere.[3–5] Numerous types of health outcomes have been investigated, and the more convincing evidence for elevated risks is generally limited to a small number of notorious waste sites. However, much of the evidence from both

---

[4] M. Vrijheid, *Environ. Health Perspect.*, 2000, **108**, 101.
[5] L. Heasman, *Waste Planning*, 1999, **3**, 3.

single-site and multi-site studies pertains to a miscellaneous assortment of hazardous waste sites (*e.g.* US National Priority List or 'Superfund' sites, which include chemical manufacturing plants, waste storage facilities, mines, contaminated groundwater areas, military bases *etc.*), illegal landfills or 'in-house' landfills within the curtailage of industry. Thus, many of the sites frequently referred to as landfills are either not landfills at all or are essentially unauthorised or unregulated chemical waste dumps that were not designed and managed for the specific purpose of waste disposal.

Whilst such sources of uncontrolled environmental pollution may have important public health implications, they are not within the scope of this chapter and are not discussed further, beyond the brief descriptions and references provided in Tables 1a and 1b for single and multi-site studies, respectively.

Review of Purported Health Outcomes

This review is limited to studies of waste sites operated for the express purpose of landfilling, thus restricting the analysis to sites conforming to the generally accepted view of what constitutes a landfill. We know of 13 single-site studies covering seven such sites[1,6-17] and an additional four relevant multiple landfill site studies.[18-22]

6 H. M. P. Fielder, C. Jones, S. R. Palmer, R. A. Lyons, S. Hillier and M. Joffe, *Report on the Study of Time to Pregnancy in the Rhondda Valleys*, Welsh Combined Centres for Public Health, Cardiff, 2000.

7 A. Mukerjee and D. Deacon, *Report on Complaints of Illhealth Perceived Due to Exposure to Nantygwyddon Landfill Site: A Descriptive Survey*, Bro Taf Health Authority, Cardiff, 1999.

8 G. Richardson, *Sarcoidosis and Nant-y-Gwyddon Landfill Site*, Bro Taf Health Authority, Cardiff, 1999.

9 M. Berry and F. Bove, *Environ. Health Perspect.*, 1997, **105**, 856.

10 M.S. Goldberg, L. Goulet, H. Riberdy and Y. Bonvalot, *Environ. Res.*,1995, **69**, 37.

11 M.S. Goldberg, N. Al-Homsi, L. Goulet and H. Riberdy, *Arch. Environ. Health*, 1995, **50**, 416.

12 M. S. Goldberg, J. Siemiatyck, R. DeWar, M. Désy and H. Riberdy, *Arch. Environ. Health*, 1999, **54**, 291.

13 M. Kharrazi, J. Von Behren, M. Smith, T. Lomas, M. Armstrong, R. Broadwin, E. Blake, B. McLaughlin, G. Worstell, and L. Goldman, *Toxicol. Ind. Health*, 1997, **13**, 299.

14 D. Zmirou, A. Deloraine, P. Saviuc, C. Tillier, A. Boucharlat and N. Maury, *Arch. Environ. Health*, 1994, **49**, 228.

15 A. Deloraine, D. Zmirou, C. Tillier, A. Boucharlat and H. Bouti, *Environ. Res.*, 1995, **68**, 124.

16 C. Hertzman, M. Hayes, J. Singer and J. Highland, *Environ. Health Perspect.*, 1987, **75**, 173.

17 K. Mallin, *Am. J. Epidemiol.*, 1990, **132**, Suppl. No. 1, S96.

18 H. M. Dolk, M. Vrijheid, B. Armstrong, L. Abramsky, F. Bianchi, E. Garne, V. Nelen, E. Robert, J. E. S. Scott, D. Stone and R. Tenconi, *Lancet*, 1998, **352**, 423.

19 M. Vrijheid, H. Dolk, B. Armstrong, L. Abramsky, F. Bianchi, I. Fazarinc, E. Garne, R. Ide, V. Nelen, E. Robert, J. E. S. Scott, D. Stone, and R. Tenconi, *Lancet*, 2002, **359**, 320.

20 P. Elliott, S. Morris, D. Briggs, C. de Hoogh, C. Hurt, T. K. Jensen, I. Maitland, A. Lewin, S. Richardson, J. Wakefield and L. Järup, *Birth Outcomes and Selected Cancers in Populations Living Near Landfill Sites, Report to the Department of Health*, Small Area Health Statistics Unit, Imperial College, London, 2001.

21 P. Elliott, D. Briggs, S. Morris, C. de Hoogh, C. Hurt, T. K. Jensen, I. Maitland, S. Richardson, J. Wakefield and L. Järup, *Br. Med. J.*, 2001, **323**, 363.

22 E. L. Lewis-Michl, L. R. Kallenbach, N.S. Geary, J. M. Melius, C. L. Ju, M. F. Orr and S. P. Forand, *Investigation of Cancer Incidence and Residence Near 38 Landfill with Soil Gas Migration Conditions, New York State, 1980–1989*, ATSDR Division of Health Studies, Atlanta, 1998.

Table 1a Single-site studies of waste sites not considered further in this review

Site	Type of site	Reason not considered	References
Stringfellow, California	Surface impoundments (waste ponds) receiving liquid industrial wastes	Not a landfill	D. B. Baker, S. Greenland, J. Mendlein and P. Harmon, *Arch. Environ. Health*, 1988, **43**, 325
Drake Superfund site, Pennsylvania	Chemical manufacture, use and storage site	Not a landfill	L. D. Budnick, D. C. Sokal, H. Falk, J. N. Logue and J. M. Fox, *Arch. Environ. Health*, 1984, **39**, 409; J. N. Logue and J. M. Fox, *Arch. Environ. Health*, 1986, **41**, 222
Hardeman County, Tennessee	Disposal site for pesticide manufacturer	Unregulated dump[a]	C. S. Clark, C. R. Meyer, P. S. Gartside, V. A. Mejeti, B. Specker, W. F. Balistreri and V. J. Elia, *Arch. Environ. Health*, 1982, **37**, 9; C. R. Meyer, *Environ. Health Perspect.*, 1983, **48**, 9
Woburn, Massachusetts	Burial pits containing animal hides and chemical wastes, and abandoned lagoons associated with industrial sites	Unauthorised or unregulated dump[a]	J. J. Cutler, G. S. Parker, S. Rosen, B. Prenney, R. Healey and G. G. Caldwell, *Public Health Rep.*, 1986, **101**, 201; S. W. Lagakos, B. J. Wessen and M. Zelen, *J. Am. Statistical Assoc.*, 1986, **81**, 583; V. S. Byers, A. S. Levin, D. M. Ozonoff and R. W. Baldwin, *Cancer Immunol. Immunother.*, 1988, **27**, 77
Sikes and French Superfund sites, Texas	Two in-house waste disposal sites for industrial wastes	Unregulated dump[a]	H. Dayal, S. Gupta, N. Trieff, D. Maierson and D. Reich, *Arch. Environ. Health*, 1995, **50**, 108
Santa Clara County, California	Leaking underground storage tank at electronics manufacturing plant	Not a landfill	M. Deane, S. H. Swan, J. A. Harris, D. M. Epstein and R. R. Neutra, *Am. J. Epidemiol.*, 1989, **129**, 894; S. H. Swan, G. Shaw, J. A. Harris and R. R. Neutra, *Am. J. Epidemiol.*, 1989, **129**, 885; M. Wrensch, S. Swan, J. Lipscomb, D. Epstein, L. Fenster, K. Claxton, P. J. Murphy, D. Shusterman and R. Neutra, *Am. J. Epidemiol.*, 1990, **131**, 283; G. M. Shaw, S. H. Swan, J. A. Harris and L. H. Malcoe, *Epidemiology*, 1990, **1**, 207; M. Deane, S. H. Swan, J. A. Harris, D. M. Epstein and R. R. Neutra, *Epidemiology*, 1992, **3**, 95; M. Wrensch, S. H. Swan, J. Lipscomb, D. M. Epstein, R. R. Neutra, and L. Fenster, *Epidemiology*, 1992, **3**, 99
Kingston, Queensland	Municipal landfill and associated unsupervised dump for disposal of chemical wastes, principally residues from oil reprocessing operation	Unauthorised dump	M. P. Dunne, P. Burnett, J. Lawton and B. Raphael, *Med. J. Aust.*, 1990, **152**, 592
Tuscon Valley, Arizona	Disposal area where contaminants placed on ground surface or in unlined earthen holding pits	Unauthorised or unregulated dump[a]	S. J. Goldberg, M. D. Lebowitz, E. J. Graver and S. Hicks, 1990, *JACC*, **16**, 155
North Rhine-Westfalia	Industrial waste dump	Unauthorised dump	E. Greiser, I. Lotz, H. Brand and H. Weber, *Am. J. Epidemiol.*, 1991, **134**, 755
Mellery Quarry	Chemical waste dump	Unauthorised dump	T. Lakhanisky, D. Bazzoni, P. Jadot, I. Joris, C. Laurent, M. Ottogali, A. Pays, C. Planard, Y. Ros and C. Vleminckx, *Mutation Research*, 1993, **319**, 317

Location	Type	Description	References
Londonderry Township, Pennsylvania	Unauthorised dump	Unauthorised dump containing drums of toxic chemicals and fly ash	W. Klemans, C. Vleminckx, L. Schriewer, I. Joris, N. Lijsen, A. Maes, M. Ottogali, A. Pays, C. Planard, G. Rigaux, Y. Ros, M. Vande Riviere, J. Vandenvelde, L. Verschaeve, P. Deplaen and T. Lakhanisky, *Mutation Res*, 1995, **342**, 147
Cambuslang, Carmyle and Rutherglen area, Glasgow, Scotland	Unregulated dump[a]	Contaminated area with six landfills associated with manufacture of chromium salts for the tanning industry	J.N. Logue, R. M. Stroman, D. Reid, C. W. Hayes and K. Sivarajah, *Arch. Environ. Health*, 1985, **40**, 155 P. McCarron, I. Harvey, R. Brogan and T.J. Peters, *Brit. Med. J.*, 2000, **320**, 11
Two Superfund sites, Harris County, Texas	Unregulated dump[a]	Two chemical manufacturing, recovery, treatment and storage sites	M.S. Miller and M. A. McGeehin, *Toxicol. Ind. Health*, 1997, **13**, 311
Walsall, West Midlands	Waste treatment site	Treatment and polymerisation site for stabilisation of toxic wastes	K. R. Muir, J.P. Hill, S. E. Parkes, A. H. Cameron and J.R. Mann, *Paediatric Perinatal Epidemiol.*, 1990, **4**, 484
Wayne Township, New Jersey	Radioactive waste site	Thorium waste disposal site	G. Reza Najem and L.K. Voyce, *AJPH*, 1990, **80**, 478
Superfund site, Southern New Jersey	Unregulated dump[a]	Disposal and storage site for chemical wastes from manufacturing plant	G. Reza Najem, T. Strunck and M. Feuerman, *Am. J. Prev. Med.*, 1994, **10**, 151
Superfund site, Galena, Kansas	Not a landfill	Heavy metals mining waste site	J.S. Neuberger, M. Mulhall, M.C. Pomatto, J. Sheverbush and R.S. Hassanein, *Sci. Total Environ.*, 1990, **94**, 261
Lowell, Massachusetts	Unregulated dump[a]	Chemical waste storage and recycling site	D. Ozonoff, M.E. Colten, A. Cupples, T. Heeren, A. Schatzkin, T. Mangione, M. Dresner and T. Colton, *Am. J. Ind. Med.*, 1987, **11**, 581
McColl waste disposal site, California	Unregulated dump[a]	Disposal site for acidic refinery sludge (by-products from aviation fuel production) and drilling mud from oil exploration	J. A. Lipscomb, L. R. Goldman, K. P. Satin, D. F. Smith, W. A. Vance and R. R. Neutra, *Environ. Health Perspect.*, 1991, **94**, 15
Love Canal, New York	Unregulated dump[a]	Disposal site for liquid and solid residues from organic pesticide production company	B. Paigen, L. R. Goldman, J. H. Highland, M. M. Magnant and A. T. Steegman, *Hazardous Waste & Hazardous Materials*, 1985, **2**, 23 D. T. Janerich, W. S. Burnett, G. Feck, M. Hoff, P. Nasca, A. P. Polednak, P. Greenwald and N. Vianna, *Science*, 1981, **212**, 1404 C. W. Heath, M. R. Nadel, M. M. Zack, A. T. L. Chen, M. A. Bender and R. J. Preston, *J. Am. Med. Assoc.*, 1984, **251**, 1437 B. Paigen, L. R. Goldman, M. M. Magnant, J. H. Highland, and A. T. Steegmann, *Human Biol.*, 1987, **59**, 489 N.J. Vianna and A.K. Polan, *Science*, 1984, **226**, 1217 L. R. Goldman, B. Paigen, M. M. Magnant and J. H. Highland, *Hazardous Waste & Hazardous Materials*, 1985, **2**, 209

[a] Unregulated dumps refer to in-house waste sites under the curtailage of industry.

Table 1b Multi-site studies of waste sites not considered further in this review

Reference	Study parameters	Type of sites evaluated[a]
L. A. Croen, G. M. Shaw, L. Sanbonmatsu, S. Selvin and P. A. Buffler, *Epidemiology*, 1997, **8**, 347	764 sites in California, USA	Inactive hazardous waste sites on US National Priority List and California EPA list, including inactive pesticide and chemical manufacturing plants, wood treatment and preserving facilities, drum storage facilities, mines, contaminated groundwater areas, sanitary landfills and military bases
S. A. Geschwind, J. A. J. Stolwijk, M. Bracken, E. Fitzgerald, A. Stark, C. Olsen and J. Melius, *Am. J. Epidemiol.*, 1992, **135**, 1197	590 sites in 20 counties in New York State, USA	Hazardous waste sites selected from NY State Hazardous Waste Site Inspection Program
J. Griffith, R. C. Duncan, W. B. Riggan and A. C. Pellom, *Arch. Environ. Health*, 1989, **44**, 69	593 sites throughout USA	Hazardous waste sites on US National Priority List with analytical evidence of contaminated groundwater providing sole source water supply
H. I. Hall, W. E. Kaye, L. S. Gensburg and E. G. Marshall, *Environ. Health*, 1996, **59**, 17	317 sites in 20 counties in New York State, USA	Sites on NY Inactive Hazardous Waste Site Registry (*i.e.* sites on US National Priority List and additional State-designated sites)
E. G. Marshall, L. J. Gensburg, D. A. Deres, N. S. Geary, and M. R. Cayo, *Arch. Environ. Health*, 1997, **52**, 416	643 sites in 18 urban counties in New York State, USA	Inactive hazardous waste sites on US National Priority List and on New York State designation list
G. Reza Najem, I. S. Thind, M. A. Lavenhar and D. B. Louria, *Int. J. Epidemiol.*, 1983, **12**, 276	Numerous sites in 21 counties in New Jersey, USA	Chemical toxic waste disposal sites listed by New Jersey State and local environmental health departments
A. P. Polednak and D. T. Janerich, *Environ. Res.*, 1989, **48**, 29	12 sites in Niagara County, New York State, USA	Inactive hazardous waste disposal sites where known or suspected lung carcinogens identified
G. M. Shaw, J. Schulman, J. D. Frisch, S. K. Cummins and J. A. Harris, *Arch. Environ. Health*, 1992, **47**, 147	300 sites in San Francisco Bay, USA	Sites with environmental contamination including landfills, chemical dumps, abandoned materials, industrial sites and hazardous treatment and storage facilities
W. A. Sosniak, W. E. Kaye and T. M. Gomez, *Arch. Environ. Health*, 1994, **49**, 251	1281 sites throughout USA	Hazardous waste sites on US National Priority List

[a]All studies examined a miscellaneous assortment of sites that were not limited to landfills and were therefore excluded from further review.

The parameters and findings of the relevant single-site and multi-site studies are set out in Tables 2a and 2b, respectively. The single-site studies examined a wide range of health outcomes and, with few exceptions, the study sites share the following characteristics: they are or were large to massive; they opened prior to the establishment of the systems of environmental control currently required in the country of concern; and they received hazardous and/or liquid chemical wastes. Approximately half of the sites closed before or during the period under study. The primary exposure pathway of concern was inhalation at six of the sites, generally of volatile organic compounds associated with landfill gas, and drinking water at one of the sites.

Three of the four multi-site studies are UK or European landfill studies that have recently received much attention in the media. They are referred to as the two EUROHAZCON studies[18,19] and the SAHSU (Small Area Health Statistics Unit) study.[20,21] These studies primarily, but not exclusively, focused on risks of congenital anomalies (birth defects) and other adverse birth outcomes such as low birth weight. The two EUROHAZCON studies examined over 20 hazardous waste landfills, the majority operating in the 1970s and 1980s. The SAHSU study examined all UK landfills operating between 1982 and 1997, including inert, semi-inert, domestic and hazardous waste landfills. The above three studies are discussed in detail in Section 3. The remaining multi-site study examined cancer incidence rates amongst populations surrounding 38 municipal landfills with soil gas migration conditions in New York State.[22] Like the sites examined by the single-site studies and by the EUROHAZCON studies, most of the New York State study landfills operated before landfills were strictly regulated, and the majority of the old landfills were not capped or lined as they would be if constructed today.

The findings from all the single- and multi-site studies have been compiled and the collective results for each investigated health outcome summarised in Table 3. A wide range of health outcomes has been investigated, although the majority of studies have examined adverse birth outcomes or cancer.

The results of these studies are inconsistent, with some reporting significant positive associations between certain health outcomes and landfill sites and others reporting no association. On balance, for each health outcome investigated, there are a larger number of studies reporting no associations than those reporting positive associations. Where associations are reported, there is little consistency in the specific type of health outcome recorded. A limited review of the broader set of epidemiological studies encompassing miscellaneous hazardous waste sites, illegal and 'in-house' landfills gave similar results, although an even wider range of health outcomes was investigated.

It is possible that, in some cases, no associations were detected due to limitations in the study design. For example, small sample sizes for single-site studies may have resulted in insufficient statistical power to detect small differences. Certain cancer studies may have been of insufficient duration to allow for the time for disease to manifest itself in an exposed population. However, the majority of studies reported significant associations for some of the specific outcomes investigated and no associations for others (Table 2), suggesting that the negative findings did not tend to be due to limitations inherent in the studies.

Table 2a Single-site epidemiological studies of potential health effects associated with landfill sites

Site	Type of waste received/years of operation/other site details	Primary exposure route/chemicals of concern	Reason for initiation of study and/or site closure	Area or population treated as being exposed/study period	Health effects examined	Presence of association with exposure
Nant-y-Gwyddon landfill, Wales (Fielder et al.,[1,6] Mukerjee and Deacon;[7] Richardson)[8]	• Household; industrial; commercial; difficult • 1988–present	Landfill gas	Community concerns that odours from site causing a variety of conditions	Fielder et al.[1] • Residents in 5 electoral wards within 3 km of site • 1981–1997	Mortality: all causes; respiratory disease; cancers	Non-significant
					Hospital admissions: general admissions; respiratory disease; asthma; cancer; sarcoidosis; spontaneous abortions	Non-significant
					Low birth weight	Non-significant
					Birth defects: all anomalies	Significant positive
					abdominal wall (gastroschisis)	Significant positive
					Drug prescription rates for gastrointestinal, respiratory and central nervous systems, skin and eyes	Elevated
				Fielder et al.[6] • Residents in 5 electoral wards within 3 km of site • 1998–2000	Time to pregnancy	Non-significant
				Mukerjee and Deacon[7] • Residents within 1 km, 1–2 km, 2–3 km and >3 km from site • 1998	Self reported symptoms: headache, sore throat, runny nose, feeling sick, diarrhoea	Elevated
					Self reported symptoms: sore eyes, dizziness, skin rash	Non-significant
					Self-reported chronic diseases	Non-significant
					Frequency of GP consultations	Non-significant
				Richardson[8] • Residents in 5 electoral wards within 3 km of site • 1991–1998	Sarcoidosis	Elevated
Lipari landfill, New Jersey (Berry and Bove)[9]	• Municipal; household; liquid and semi-solid chemical; other industrial • 1958–1971	Inhalation of volatilised chemicals emitted from landfill and from contaminated waters	Public complaints regarding odour, respiratory problems, headaches, nausea and dying vegetation	• Radius of 1 km from perimeter of site, including high exposure group adjacent and downwind of site • 1961–1985	Average birth weight	Significant positive
					Proportion low birth weight	Significant positive
					Proportion premature births	Significant positive

Site / landfill	Exposure	Concerns	Study details	Outcome	Association
• Ranked no. 1 on US EPA's National Priority List • Liquid wastes emptied from containers prior to disposal • Hazardous leachate migrated into nearby streams and a lake immediately adjacent to community with homes, schools and playgrounds					
Miron Quarry, Quebec; (Goldberg et al.)[10,11,12] • Domestic; industrial; commercial • 1968–present • 3rd largest municipal solid waste landfill site in North America • 100 000 people live within 2 km • has not been capped • biogas collection system installed in 1980, and operated at low efficiency	Release of landfill gas into ambient air and soil	Health concerns expressed by local residents; frequent odour complaints registered	Goldberg et al.[10] • Postal code areas containing and bordering site (up to 4km from perimeter of site) • 1979–1989	Low birth weight Very low birth weight Small for gestational age Preterm births	Significant positive Non-significant Significant positive Non-significant
			Goldberg et al.[11] • Postal code areas containing and bordering site (up to 4km from perimeter of site) • 1981–1988	Males: cancers of stomach; liver and intrahepatic bile duct; trachea, bronchus and lung; prostate Females: cancer of stomach; cervix uteri Females, breast 13 other cancer sites in males; 17 other cancer sites in females	Significant or nearly significant positive Significant or nearly significant positive Significant negative No association
			Goldberg et al.[13] • Postal code areas containing and bordering site (up to 4km from perimeter of site) • 1979–1985	Males: cancer of liver; kidney; pancreas; prostate; and non-Hodgkin's lymphomas 8 other cancer sites in males	Significant or nearly significant positive No association
BKK landfill, California (Kharrazi et al.)[13] • Hazardous waste of all types; municipal • 1963–1989 • Received nearly 4 million tons of hazardous waste • Residential developments in close proximity • Numerous complaints of odour, surface water run-off onto nearby streets, hazardous waste spills from HGVs, and dust releases	Airborne exposures	Concerns over public health and welfare following complaints of odours, surface water run-off, hazardous waste spills from trucks and dust releases	• Residence in areas with high rates of odour complaints (high odour area up to 0.6 miles from landfill) • 1978–1986	Reduction in gestational age Low mean birth weight Fetal and infant mortality	Significant positive Significant positive No association

Table 2a (*cont.*)

Site	Type of waste received/years of operation/other site details	Primary exposure route/chemicals of concern	Reason for initiation of study and/or site closure	Area or population treated as being exposed/study period	Health effects examined	Presence of association with exposure
Montchanin landfill, France (Zmirou et al.,[14] Deloraine et al.)[15]	• Liquid and solid toxic industrial, including wastewater treatment sludge, dehydrated hydroxide sludge and solvent-containing wastes • 1979–1988 • Received 400000 tons of industrial wastes • Located adjacent to town of 6000 inhabitants – 100 m from nearest houses	VOCs in ambient air	Community health concerns triggered by offensive odours, suspected increase in certain health complaints, and elevated levels of VOCs in ambient air	Zmirou et al.[14] • Estimated exposures using air dispersion model • 1987–1989 Deloraine et al.[15] • Estimated exposures using air dispersion model • 1990	Drug consumption rates for respiratory, opthalmological, dermatological, gastro-intestinal and neurological conditions Psychiatric disorders Respiratory symptoms Isolated biological abnormalities Skin diseasese Eye diseases Ear, nose and throat conditions Miscellaneous conditions	No significant association Significant positive Significant positive No association No association No association No association No association
Upper Ottawa Street landfill, Ontario (Hertzman et al.)[16]	• Solid and liquid industrial; commercial; domestic • 1950s–1980 • Volumes of industrial waste received increased throughout 1970s, such that approx. 8–12 million gallons of liquid waste disposed of during 1978 • Capped in 1980/81	Airborne exposures to vapours, fumes, dust or ash, as well as direct skin contact	Public concerns regarding health effects	• Residence within 750 m from edge of tipping face • Approx. 1984	Self-reported respiratory, skin, mood, narcotic and eye conditions Self-reported muscle weakness Self-reported adverse birth outcomes: low birth weight stillbirth miscarriage/spontaneous abortion birth defects	Significant positive No association No association No association No association No association
Waste disposal site, Northwestern Illinois (Mallin)[17]	• Municipal; industrial, including solvents, plating wastes and heavy metals • Late 1950s–1972	Drinking water from wells contaminated with VOCs	Several areas of elevated mortality from bladder cancer identified in region	• Residence in town using water from contaminated wells • 1977–1985	Bladder incidence	Significant positive

Table 2b Multi-site epidemiological studies of potential health effects associated with waste disposal sites

Author(s)	Study Parameters	Type of sites evaluated/years of operation	Area or population treated as being exposed/study period	Health effects examined	Presence of association with exposure
Dolk et al.[18]	21 landfill sites in 5 European Countries	• Landfill sites handling hazardous chemical wastes • Majority either opened before mid-1970s or closed before mid- to late-1980s	• Maternal residence within 3 km of landfill site • Mid/late 1980s–1993 in most cases	Non-chromosomal birth defects: all anomalies; neural tube; cardiac septa; great arteries and veins	Significant positive
				Non-chromosomal birth defects: tracheo-oesophageal; hypospadias; gastroschisis	Nearly significant positive
				19 other specified types of non-chromosomal birth defects	No association
Vrijheid et al.[19]	23 landfill sites in 5 European Countries	• Landfill sites handling hazardous chemical wastes • Majority either opened before mid-1970s or closed before mid- to late-1980s	• Maternal residence within 3 km of landfill site • Mid/late 1980s–1993 in most cases	Chromosomal birth defects	Significant positive
Elliott et al.[20,21]	9565 landfill sites in England, Wales and Scotland	• 774 special waste landfills, 7803 non-special waste landfills, and 988 classified as unknown • Sites operational between 1982 and 1997	• Residence within 2 km of landfill site • 1983–1998	Birth defects: all anomalies; neural tube; hypospadias/epispadias; abdominal wall; gastroschisis/exomphalos	Excess risks
				Birth defects: cardiovascular	Depressed risks
				Low and very low birth weight	Excess risks
				Still births	No association
				Cancer registrations: bladder; brain; hepatobiliary; childhood and adult leukaemia	No association
Lewis-Michl et al.[22]	38 landfill sites in New York State, USA	• Municipal landfills with soil-gas migration conditions; selected from the New York State Inactive Hazardous Waste Site Registry; sites in NY City excluded • Majority of landfills opened prior to 1970, closed prior to end of 1980s • Majority not capped or lined	• Residence within 250 ft of landfill site boundary (or greater distance if further gas migration shown) • 1980–1989	Male cancer incidence: liver; lung; bladder; kidney; brain; non-Hodgkin's lymphoma; leukaemia Female cancer incidence: liver; lung; kidney; brain; non-Hodgkin's lymphoma	No association
				Female cancer incidence: bladder; leukaemia	Significant positive

Table 3 Summary of findings of epidemiological studies at landfill sites

Health outcome	Number of studies indicating excess risks	Number of studies indicating no excess risk
Birth defects:		
All chromosomal anomalies	1	0
All non-chromosomal anomalies	3 (before site opened in 2 of these)	1
Central nervous system	0	1
Neural tube defects	2	0
Cleft lip/palate	0	1
Heart and circulatory	1	1
Hypospadias/epispadias	2 (borderline in 1; before site opened in 1)	0
Limb reductions	0	1
Abdominal/gastroschisis	3 (borderline in 1; before site opened in 1)	0
Skin and other integument	0	1
Tracheo-oesophageal	1 (borderline)	0
Renal	0	1
Urinary tract	0	1
Other pregnancy outcomes:		
Low birth weight/prematurity	4	2
Still births	0	3
Infant mortality	0	1
Spontaneous abortions	0	2
Time to pregnancy	0	1
Cancer:		
All types	0	1
Oesophagus	0	1
Stomach	1	1
Liver	2	2
Trachea/bronchus/lung	1	2
Prostate	2	0
Cervix uteri	1	0
Breast	0	1
Colorectum	0	2
Brain	0	3
Pancreas	1	1
Kidney	1	2
Bladder	2	3
Leukaemia	1	2
Non-Hodgkin's lymphoma	1	2
Skin melanoma	0	1
Respiratory		
All respiratory diseases	1	1
Asthma	0	1
Sarcoidosis	1 (before industrial tipping commenced)	1
Psychiatric disorders	1	0
Miscellaneous self-reported symptoms	2	0
Drug prescription rates, miscellaneous symptoms	1	1
Unspecified:		
Hospital admissions, all diseases	0	1
Mortality, all causes	0	1

114

Both of the studies examining miscellaneous self-reported symptoms reported significant excess risks in the vicinity of the respective landfill site. Residents living closer to the Nant-y-Gwyddon landfill site complained of more frequent symptoms, including headache, sore throat, runny nose sickness and diarrhoea.[7] Residents in the immediate vicinity of the Upper Ottawa Street landfill, Ontario reported excess respiratory, skin, mood, narcotic and eye conditions.[16] The results of studies of self-reported symptoms should be interpreted with caution as they are prone to recall and reporting bias or may be explained by concern and worry about living near the landfill rather than the direct effects of chemical releases.

On face value, the balance appears to be towards a significant adverse effect for total birth defects, neural tube defects, hypospadias/epispadias and abdominal wall defects, each showing apparent associations in two or three studies. However, total birth defects showed excess risks prior to commencement of landfill operations in two of these studies.[1,6,20,21] Hypospadias/epispadias and abdominal wall defects were elevated prior to commencement of operations in one of the studies[20,21] and risks of borderline significance were shown in a second study.[18] Thus, on closer inspection there is less consistency in the results for these outcomes than it would at first appear.

The only other health outcome that exhibits excess risks with any consistency is low birth weight. Low birth weight was reported as being in significant excess by four of seven studies.[9,10,13,20,21] Low birth weight is associated with a large number of risk factors, including smoking, socio-economic status, parental height and nutritional factors, which may give rise to misleading apparent links between body weight and landfill exposure.[4] The epidemiological studies on birth weights attempted to account for such potential confounding factors, but it is not known whether all relevant factors were successfully accounted for.

Conclusion

There is a growing body of epidemiological literature examining health risk in the vicinity of hazardous waste sites, although studies focusing on authorised landfill sites operated by qualified waste management companies are more scarce. The landfill sites that have been investigated tend to be old, large, subject to limited environmental controls, and to have received hazardous and/or liquid chemical wastes. The primary exposure pathway of concern at the majority of reviewed sites was inhalation, generally of volatile organic compounds associated with landfill gas. The collective findings from the landfill studies indicated that, on balance, there are a larger number of studies reporting no associations than those reporting positive associations and there is little consistency in terms of individual specific effects. The health outcomes coming closest to consistently showing effects were birth defects, which frequently exhibited apparent elevated risks near landfills before the sites opened, and low birth weight, which is known to be associated with a number of factors that may have a confounding influence on incidence rates in the vicinity of landfills. A number of studies reported increased risks of various self-reported symptoms; these may have resulted from recall and reporting bias or anxiety associated with living near a landfill rather than the direct effects of chemical releases.

3 Adverse Birth Outcomes

This section takes a more detailed look at all the landfill studies reporting excess risks of birth defects and low birth weight as these are the health outcomes that have been rightly or wrongly most consistently linked with landfills. The review focuses on four recent studies of UK and European landfills, as they provide the bulk of the evidence on the risks of adverse birth outcomes near landfills, they have been the subject of much scientific debate, and they illustrate some of the difficulties associated with undertaking epidemiological studies. The studies include one single-site study on the Nanty-y-Gwyddon Landfill in Wales,[1] the two EUROHAZCON studies[18,19] and the SAHSU study.[20,21]

We also refer to three additional studies reporting increased risk of low birth weight in the vicinity of North American landfills: the Lipari Landfill in New Jersey,[9] Miron Quarry in Montreal,[10] and the BKK Landfill near Los Angeles.[13]

The above studies are discussed in terms of whether they provide robust evidence for a significant statistical link between landfill and adverse birth outcomes. This section focuses on birth outcome studies at landfills with particular reference to the following methodological issues: (1) identifying appropriate reference (unexposed) areas, (2) accounting for alternative sources and other confounding factors and (3) accounting for random chance effects and other statistical artefacts associated with rare, highly variable health outcomes.

Nant-y-Gwyddon Study

Fielder *et al.*'s single-site study of the Nant-y-Gwyddon landfill site reported evidence for a possible link between maternal residential proximity and birth defects, but no evidence for a link with low birth weight.[1] It was commissioned as a result of an increase in the number of local complaints regarding odours and health concerns. In particular, there had been concerns about increased rates of gastroschisis, an anomaly of the abdominal wall. The study was a geographical comparison of health effects within five 'exposed' wards within 3 km of the landfill and from which formal complaints of odours had been made, and in 22 other unexposed reference wards of similar socio-economic status in the same local authority.

The study found statistically significant elevated rates of reported birth defects in the exposed wards between 1983 and 1996. This increase occurred both before and after the opening of the landfill site in 1988. A cluster of four babies born with gastroschisis was confirmed in the exposed wards from 1991, post-dating the opening of the landfill. A separate analysis for all birth defects occurring in 1998 indicated that the rate for the exposed wards was slightly higher than that for the unexposed wards, but no higher than the rate for Wales overall.[23]

Fielder *et al.*'s stated objective was '*To compare indices of health in a population living near a landfill site with a population matched for socioeconomic status...*'[1] Whilst it would seem to be a scientific requisite that the matched population should not be exposed to potential risks from landfills, Fielder *et al.* provide no indication that they have attempted to exclude areas containing such landfills

---

[23] J. Greenacre, M. Morgan and D. Tucker, *Br. Med. J.*, 2000, **320**, 1542.

from their reference wards. It has been previously suggested that site licence records indicate that there are a number of other operating or closed landfills whose zones of potential influence include the 22 reference wards, including sites which, like the Nant-y-Gwyddon landfill, may have been permitted to receive difficult wastes.[3,24]

The presence of these additional landfills in reference wards leads to irresolvable interpretational difficulties: the interpretation of Fielder *et al.*'s data depends on the relative magnitudes of exposures and hazards, if any, associated with the respective landfills. If it is assumed that the additional landfills represent some degree of exposure and hazard, but that these are lower in magnitude than the Nant-y-Gwyddon landfill, then this would suggest that Fielder *et al.* have under-estimated the risks associated with Nant-y-Gwyddon. Alternatively, it may be concluded that the additional landfills are not having any measurable health effects and that the reported health effects are unrelated to landfill sites in general but specific to the Nant-y-Gwyddon site. Finally, there may be no connection between the reported health effects and landfill operations at all. The latter appears more likely, considering the occurrence of elevated incidence rates prior to commencement of operations at the Nant-y-Gwyddon landfill.

Given these considerations, it seems reasonable to pose the question as to whether there might be other causal factors that would be equally or more likely to account for the elevated birth rates than the presence of the landfill. However, the authors of the study made no reference to any attempts to identify other potential sources in the study area, although at least one is known to have occurred, a former municipal waste incinerator located about 2 km from Fielder *et al.*'s exposed wards.[3,24] Since Fielder *et al.* provide no direct evidence that the Nant-y-Gwyddon landfill is the cause of the reported elevated rates of birth defects, it could equally be asserted that the incinerator is a material factor, although there may be other causal factors, including environmental and genetic. Limited ambient air exposure measurements taken in the community surrounding the Nant-y-Gwyddon site provided no clear indication of the likely cause of the elevated risks or the confirmed gastroschisis cluster.

The study appears to have identified an area of land with high incidences of birth defects, although the cause(s) of this are unknown.

EUROHAZCON Studies

The EUROHAZCON project has resulted in two multi-site landfill studies on birth defects and residence near hazardous waste landfill sites. Dolk *et al.* examined non-chromosomal birth defects, while Vrijheid *et al.* examined Down's syndrome and other chromosomal defects.[18,19] Non-chromosomal defects are adverse teratogenic (developmental) conditions that result from abnormal cell growth in the developing embryo or foetus. Chromosomal defects are associated with major structural mutations in the child's chromosomes resulting from abnormal cell division during the process of sperm or egg production in the parents.

Dolk *et al.* used data on non-chromosomal birth defects from seven research

---

[24] R. D. Roberts, A. Redfearn and J. C. Dockerty, *Br. Med. J.*, 2000, **320**, 1541.

centres maintaining regional population-based registers.[18] The seven centres are in five European countries. 21 hazardous waste landfills located in the regions covered by the centres were identified as being suitable for study. The area of land within a 7 km radius of each landfill was defined as the study area. Certain landfills were located within 7 km of one another, and these were combined into larger study areas such that the 21 identified landfills were grouped into a total of 15 study areas.

The study was a case-control study involving a total of 1089 birth defect cases and 2366 controls randomly selected from babies with no malformations born in the same study area on the nearest following day or in the same year. Cases and controls born to mothers living within 0–3 km of a landfill were assigned to the 'exposed' category, and those born to mothers living within a 3–7 km radius were assigned to the 'unexposed' reference category. The risk for the exposed category was determined as the Odds Ratio, the number of cases as a proportion of the number of controls relative to this proportion for the unexposed category. Odds Ratios were adjusted to account for confounding by maternal age and socio-economic status.

The significance of the risks or Odds Ratios was assessed by statistical analysis for all study areas combined (for all birth defects and for 22 specific types of anomaly) and for each individual study area. In addition, the combined data for all birth defects were grouped into six distance bands and the statistical relationship between distance from the study landfills and the Odds Ratios was examined.

The principal findings were as follows:

- For all data combined and all non-chromosomal birth defects, residence within 3 km of a landfill site was associated with a significant (33%) increase in risk;
- For all data combined and all non-chromosomal birth defects, Odds Ratios for six distance bands between 0–1 km and 5–7 km showed a consistent significant degradation in risk with increasing distance from a landfill;
- For all data combined for the 22 specific types of birth defects, risks were significantly raised within 3 km of a landfill site for anomalies of the neural tube, cardiac septa, and great arteries and veins, and marginally raised for a further three types of anomaly;
- Risks for all birth defects were significantly raised within 3 km of a landfill site for three of the individual study areas (containing nine landfills); the remaining 12 study areas (containing 12 landfills) showed either no statistically significant increase or in some areas a possibly lower frequency near the target landfill;
- The data for the three 'significant' study areas appear to be entirely responsible for the overall significance of the data combined across all 15 sites; the pooled data from the other 12 study areas showed no increase in risk of congenital anomaly associated with residence within 3 km of a landfill.

Vrijheid *et al.* used similar methodologies for data on chromosomal birth defects from the seven regional registers, as well as data from two regions within the England and Wales Down's Syndrome Register.[19] A further three hazardous landfills were identified in the two additional regions. One of the 21 landfills originally included in Dolk *et al.*'s study was excluded, leaving a total of 23 landfills in 17 study areas. The study included a total of 245 chromosomal congenital anomaly cases and 2412 controls.

The principal findings were as follows:

- For all data combined and all chromosomal birth defects, residence within 3 km of a landfill site was associated with a significant (41%) increase in risk compared to the 3–7 km category; however, risks did not decline consistently with distance, such that more pronounced increases in risks were found for the 2–3 km distance band (74%, significant) than for the 0–1 km (68%, non-significant) and 1–2 km (8%, non-significant) distance bands;
- Risks for Down's syndrome and non-Down's syndrome considered separately were elevated, but not significantly so;
- Risks for all chromosomal defects were significantly raised within 3 km of a landfill site for only one of the 17 individual study areas (containing one landfill); the remaining 16 study areas (containing 22 landfills) showed either no statistically significant increase or, in three areas, a possibly lower frequency near the target landfill.

In summary, whilst both studies reported increased overall risks for birth defects within 0–3 km compared with 3–7 km, only non-chromosomal anomalies showed a consistent decrease in risk with distance away from the sites; chromosomal birth defects did not consistently decline with distance. Significantly elevated risks were reported for all non-chromosomal defects combined, all chromosomal defects combined and three specific non-chromosomal defects. Significantly elevated risks for all non-chromosomal and chromosomal defects were reported for the data combined across all study areas and within a limited subset of the individual study areas. Because the detected risks are small, the subgroup analyses (individual study areas and individual anomalies) are associated with low statistical power due to small sample sizes.

Given the positive associations indicated for both non-chromosomal and chromosomal anomalies, the authors concluded that '*Either landfill exposures are causally related to risk of congenital anomaly and have both teratogenic and mutagenic effects, or the relation is not causal and findings indicate a common bias, or a chance effect in the selection of a common pool of control births.*'[19]

One potential common bias is the presence of other industrial sites or toxic exposures near landfill sites. As previously indicated elsewhere, it is understood that, for example, the study areas in the UK also include a number of other potential sources of chemical releases to the environment; some are located in heavily industrialised areas.[3] Potential alternative sources within the UK study areas include sewage works, Integrated Pollution Control (IPC) Part A and Part B process sites, areas of heavily contaminated land and additional landfills licensed to accept difficult or hazardous wastes.

Even if one assumes that the landfills under study are a causal factor, the results pertain only to a limited subset of landfills. There are a number of reasons for this. Firstly, the studies were limited to landfill sites that handle hazardous industrial wastes, and thus did not include municipal landfill sites which receive only domestic waste and similar types of waste from commercial and industrial premises or sites which receive only inert waste. Secondly, the data identify a potential significant effect for only a limited number of the individual landfills included in the study areas: 9 of 21 in Dolk *et al.*'s study, and 1 of 23 in Vrijheid *et al.*'s study (although the number of cases was small within individual study areas, which reduces the statistical power of the tests).

The third possible reason relates to the fact that, like the Nant-y-Gwyddon study, the EUROHAZCON studies do not appear to have adequately accounted for non-target landfills in study areas, and thus have not identified appropriate unexposed reference areas. The target landfill sites included in the studies were those containing '*hazardous waste of non-domestic origin, as defined in the EC Directive on Hazardous Waste*'.[18] It is understood that the study areas were selected such that no non-target hazardous waste sites, as defined above, occur in the 3–7 km reference areas, but that no attempt was made to ensure that municipal waste landfills do not occur in such areas. The majority of the target landfills operated in the 1970s and 1980s. At this time, landfills were not subject to the strict regulations that are in place today, and many sites received a variety of wastes, including those that would and those that would not conform to the definition of 'hazardous' in the EC Directive. Thus, the distinction between 'hazardous' and 'municipal' landfills is not always clear, and it is far from certain that any so-called 'municipal' landfills located within the 3–7 km reference areas and operating during the relevant period did not also receive significant quantities of 'hazardous' waste. In any event, all available scientific evidence indicates that emissions from hazardous landfills in the UK are no different in terms of pollutant characteristics to those from non-hazardous municipal waste landfills. The SAHSU study discussed below reported no material differences in risks for special (hazardous) waste sites and non-special waste sites.[20,21]

If there are in fact no relevant non-target landfills located in reference areas, the failure to account for municipal landfills would be merely an inconsequential weakness in the experimental design of the studies. If, however, relevant non-target landfills do occur in reference areas (and it is suspected that they do, as indicated previously elsewhere),[3] then these additional landfills introduce serious interpretational difficulties. As discussed for the Nant-y-Gwyddon study, the interpretation depends on the relative magnitudes of the exposures and hazards, if any, associated with the respective landfills. If the additional landfills have a material effect on risks, though of a lesser magnitude than the target landfills, the conclusion would be that the reported risks for the target landfills may be under-estimated. Alternatively, the additional landfills may have no material effect on risks, in which case the reported risks are confined to a certain proportion of hazardous waste landfills (*i.e.* there is a gradation of risk between landfills).[3] The final alternative explanation is that landfill *per se* is not a causal factor at all.

The EUROHAZCON studies appear to have identified certain areas of land with a higher incidence of birth defects, but the causal factor(s) – whether it be landfill, sewage works, contaminated land or other causes – has not been identified. The authors themselves stressed that their study did not establish any causal link and that far more research would be required before any firm conclusions can be drawn. Even if one assumes that landfill is a causal factor, the studies are confined to hazardous waste landfills and the data appear to identify a potential effect for a small proportion of the landfills included in the study areas.

SAHSU Study

A UK-wide epidemiological study by the Small Area Health Statistics Unit (SAHSU) into possible relationships between landfill and health was undertaken as part of the UK Government's response to earlier studies, in particular Dolk *et al.*'s EUROHAZCON study.[20,21]

The SAHSU study is particularly important for two reasons. Firstly, it encompasses all 9565 known landfills in the UK operational between 1982 and 1997 and is, as a result, the largest study of its kind ever undertaken. Secondly, the 'reference' population is confined to areas that specifically do not include landfills. The study differs in this respect from the EUROHAZCON studies and the Nant-y-Gwyddon study, which have not excluded exposure to landfill from the 'reference' population.

The primary focus of the SAHSU study was the rate of adverse birth outcomes including birth defects, low birth weight, and still births throughout the UK (although cancer incidence data were also evaluated). The health outcome data were divided, according to postcode, into two groups – those within 2 km of a known landfill (55% of the national population) and those not within 2 km (20% of the national population). (The remaining 25% of the population resided within 2 km of a landfill that closed before 1982 or opened after 1997.) Data for the second group were modelled to produce theoretical standard rates for an unexposed or reference area, including numerical adjustments for factors known to influence each health effect, such as year of birth, administrative region, sex and socio-economic deprivation. The modelled rates were then compared to the data for the 'exposed' population (*i.e.* those living within 2 km of any known landfill) and the results expressed as 'exposed' *versus* 'reference' ratios (relative risks), together with calculated 99% confidence limits. Where the upper and lower 99% limits both exceeded a ratio of 1.0 the authors inferred an increased rate of an outcome in the population living near to landfills. Similarly, where the 99% limits were both below 1.0 it can be inferred that there is a decreased rate of an outcome in the population living near to landfills, and where the 99% limits straddle 1.0, there is no difference between the two groups.

After adjustment of the data for other known influencing factors, the study outputs indicated positive associations (*i.e.* where the 'exposed' rate appears to be higher than the 'reference' area rate) for total birth defects (chromosomal and non-chromosomal), three individual types of non-chromosomal anomalies (neural tube defects, hypospadias/epispadias and abdominal wall defects) and low birth weight. No positive associations were reported for stillbirths or cancers. A negative association was indicated for cardiovascular anomalies. For landfill sites that opened during the study period, total birth defects and some specific birth defects exhibited higher risks in the period prior to opening compared with the period during operations or after closure. Results for sites known to receive special (hazardous) wastes did not differ materially from those for non-special sites.

The authors of the SAHSU study concluded that further studies are required to determine whether the findings reflect a causal mechanism or result from residual confounding or other data artefacts. There are a number of inconsistencies within the study outputs that would suggest that the results may arise from data artefacts. For example, the results show an apparent negative association with landfill for total birth defects for their unadjusted data (mean ratio of 0.92 and 99% confidence intervals of 0.907–0.923) whereas the adjusted data appear to show a small positive association (mean ratio of 1.01 and 99% confidence intervals of 1.005–1.023). Conversion of a negative into a positive association as a result of adjustment for confounding factors is unusual in epidemiological studies.

SAHSU's results for cardiovascular defects indicate an apparent negative association for both unadjusted and adjusted data. It is difficult to envisage a source–pathway–target exposure mechanism that could give rise to a reduction in the rate of cardiovascular defects in populations living near to landfills. In reality, the apparent negative association is likely to be a statistical artefact. However, where such an effect occurs, it naturally raises the question as to whether apparent differences in the opposite direction reflect real increases or result from residual confounding or statistical artefacts.

Landfill sites in the UK, and thus the 'exposed' areas in the SAHSU study, are more likely to be located in highly populated urban areas (or at least at the fringes of urban areas).[25] Urban areas tend to have higher levels of air pollution as a result of industrial activities and traffic. Socially deprived areas tend to have higher adverse birth outcome rates than affluent areas. Thus, even if landfill *per se* does not cause adverse health effects, one might expect apparent statistical associations because they tend to be located in highly industrial, polluted or socially-deprived areas.

Such statistical confounding and other artefacts may be teased out only through detailed forensic investigation and sensitivity analyses of the data. The SAHSU study includes such analyses for a number of the potential confounding factors. As discussed above, one of the potential confounding factors is urban:rural status. Reference areas in the SAHSU study are in a relatively high proportion of rural locations, predominantly in unpopulated areas of Scotland and Wales which are generally uninfluenced by the environmental effects of urbanisation and which may have a narrower genetic pool than that encountered in SAHSU's 'exposure' grouping. Whilst the SAHSU study has attempted to adjust its risk estimates for these and other confounding factors, these adjustments were based only on data for reference areas.

SAHSU also examined the effects of landfill in rural areas only, and found diminished risks for neural tube defects (adjusted relative risk of 0.99 for rural areas only compared with 1.05 for rural and urban areas combined) and hypospadias/epispadias (1.01 compared with 1.07). 99% confidence intervals on the risk estimates straddled 1 for all birth defects and all specific anomalies, indicating no landfill effect, although this may reflect the smaller sample size and resulting lack of precision.

We have undertaken an analysis of SAHSU's birth defect data for the combined study population (reference and exposed) for urban *versus* rural areas. By combining data from several of SAHSU's tables,[20] we were able to derive

---

[25] D. Briggs, K. de Hoogh, C. Hurt and I. Maitland, *A Geographical Analysis of Populations Around Landfill Sites*, Small Area Health Statistics Unit, Imperial College, London, 2001.

relative risk estimates for urban *versus* rural areas for the entire study population. The two right-hand-most columns in Table 4 show these relative risk estimates for urban *versus* rural status as well as relative risk estimates for landfill (*i.e.* for exposed *versus* reference populations). Table 4 indicates that risks for urban areas were generally elevated and correlated quite closely with SAHSU's reported risks for landfill exposures (adjusted for year, region and deprivation). The urban:rural ratios are sufficient to account for the landfill ratios. Thus, SAHSU's results are consistent with the confounding influence of urban:rural status, which may not have been adequately accounted for in the SAHSU study (through adjusting for the effect using reference area data only). This would also provide one logical explanation for why cardiovascular defects, the only type of anomaly that occurred more frequently in rural and reference areas, showed an apparent negative association with landfill exposures. Whether or not this explanation is correct, this analysis at least illustrates the kind of forensic investigation that should be applied to data susceptible to artefacts.

Birth defect data are, by their very nature, susceptible to statistical artefacts: they are characterised by rates that are small in magnitude, but large in variation. Data with large coefficients of variation are susceptible to the detection of apparent spatial patterns (or temporal differences, as in SAHSU's comparisons of areas before and after landfills opened) as a result of random chance or the effects of variables other than those under study. The fact that some specific birth defects exhibited higher risks in the period prior to opening compared with the period during operations or after closure suggests that factors independent of landfill may be at work.

A further implication of the variability in birth defect data is that the magnitude of any apparent increase needs to be placed in a broader context. Table 5 illustrates the relative magnitude of the variation in unadjusted incidence rates due to the purported effects of landfill compared to the variation due to differences amongst the 10 administrative regions in Great Britain. The apparent increases in risks of birth defects reported in the SAHSU study for areas near landfill generally range between 1 and 10% with occasional increases of up to 26%. The incidence rates for reference areas amongst the nine administrative regions in England and Wales span a range of between 19% (for all birth defects) and 107% (for abdominal wall defects), or, if one includes Scotland where reported rates were consistently high, between 147 and 1072%. The differences amongst the regions may partly reflect differences in reporting procedures or confounding factors such as socio-economic status. However, given the scale of the differences in apparent risks between one region and another, the quantum of purported increased risks associated with landfill seems to be of relatively minor magnitude, and hence, it is difficult to ascribe any differences to landfill with any degree of confidence.

Other Studies Reporting Low Birth Weights

Berry and Bove compared birth weights in babies born to parents living closest to the Lipari Landfill in New Jersey with those born to parents living within the four nearest towns at distances greater than 1 km.[9] The Lipari site received liquid and

Table 4 Relative risks of birth defects for rural vs. urban status and for exposed vs. reference areas (data based on SAHSU study)

	Urban areas			Rural areas			Relative risk (urban vs. rural)	Relative Risk (exposed vs. reference – adjusted)
	No. of cases	No. of births	Rate (per 100 000 births)	No. of cases	No. of births	Rate (per 100 000 births)		
All birth defects	99 447	6 235 739	1595	25 150	1 615 910	1556	1.02	1.01
Neural tube defects	3729	6 235 739	60	919	1 615 910	57	1.05	1.05
Cardiovascular defects	7425	6 235 739	119	2014	1 615 910	125	0.96	0.96
Hypospadias and epispadias	7966	3 192 646	250	1882	828 637	227	1.10	1.07
Abdominal wall defects	1573	6 235 739	25	363	1 615 910	22	1.12	1.08

Table 5 Variation in unadjusted congenital anomaly incidence rates amongst different administrative regions (reference areas only) and between reference and exposed areas. (Data based on SAHSU study)

	Effect of administrative region			Effect of landfill		
	Range for England and Wales (rate per 100 000 births)	Scotland (rate per 100 000 births)	%age Difference for E & W (and Scotland)	Reference (rate per 100 000 births)	Exposed (rate per 100 000 births)	%age difference (unadjusted)
Register and terminations						
All birth defects	1230–1469	7516	19 (511)	1694	1550	−8
Neural tube defects	41–76	104	85 (154)	56	60	+7
Cardiovascular defects	72–141	844	96 (1072)	134	115	−14
Hypospadias and epispadias	191–299	569	57 (198)	240	247	+3
Abdominal wall defects	15–31	37	107 (147)	22	26	+16
Hospital admissions						
Hypospadias and epispadias (surgical corrections)	–	–	–	268	257	−4
Abdominal wall defects	–	–	–	35	40	+13
Gastroschisis and exomphalos (surgical corrections)	–	–	–	19	25	+26

semi-solid chemical wastes, other industrial wastes, municipal and household wastes between 1958 and 1971, and was ranked number one on the US EPA's National Priority List. The site was closed in 1971 because of residents' complaints regarding odours and health concerns. The primary pathways for public exposures were inhalation of volatilised chemicals emitted from the landfill and from lakes and streams that had been contaminated by leachate. It was postulated that the period of highest exposures was 1971–1975 because the heaviest migration of pollution was thought to have occurred immediately prior to and during this period.

Having controlled for a number of influencing factors such as mother's age, education and the level of prenatal care, average birth weights were significantly depressed and the proportion of low weight births was significantly elevated in the population adjacent to the landfill and contaminated lake during 1971–1975. Risks of prematurity were significantly elevated in the exposed population during the 1971–1975 period. It was concluded that unmeasured risk factors such as alcohol consumption and smoking were not playing a role because the adjacent population had significantly elevated average birth weights during the early period of landfill operation (1961–1965) and during the 5 year period following the highest exposure period. Whilst a large number of volatile organic compounds were detected off-site, including chemicals with potential embryotoxic or fetotoxic effects, it is not known whether these occurred at levels in ambient air sufficient to cause the reported effects, or how levels varied throughout the duration of the study. No information was provided on potential alternative pollutant sources in the study area.

Goldberg *et al.* examined risks of adverse reproductive outcomes, including low birth weight, in the vicinity of the Miron Quarry municipal solid waste landfill site in the heart of Montreal, Quebec.[10] The Miron Quarry is the third largest municipal solid waste landfill site in North America, having received approximately 36 million tons of domestic, commercial and industrial wastes between 1968 and the time of the report. Landfill gas emissions are the principal environmental and health concern at the site. The landfill has not been capped, and the gas abstraction, collection and combustion system installed in 1980 has operated inefficiently.

Potential exposure to landfill gas was defined using a number of exposure zones consisting of postal code areas. The high exposure zone was defined by the postal code areas containing and bordering the site, and ranged between 2 km and 4 km from the edge of the site. Reference zones from subsets of unexposed areas of Montreal were selected to be similar to the exposure zones with respect to several key socio-demographic factors. After adjustment for further potential influencing factors, low birth weight and small-for-gestational-age were significantly or nearly significantly elevated in the high exposure zone compared with the reference zone. Risks for small-for-gestational-age were more elevated in the medium exposure zone than in the high exposure zone, suggesting that the apparent risks may be explained by factors other than landfill exposures. No significant associations were observed for very low birth weight and prematurity. Because it was not possible to account for all potential influencing factors such as maternal tobacco smoking and medical conditions during pregnancy, and because detailed exposure assessments were not available, the authors were not able to definitively conclude whether the reported effects were due to landfill gas exposures. No information was provided on the presence of alternative sources of environmental pollution in the study area.

Kharrazi *et al.* examined adverse pregnancy outcomes, including low birth weights and shortened gestational periods in the vicinity of the BKK Landfill near Los Angeles.[13] The site received municipal waste from 1963 and, in a separate on-site facility, nearly 4 million tons of commercial hazardous waste of all types from before 1972 until 1984. During the period of operations, residential areas expanded into the immediate vicinity of the landfill, and numerous complaints relating to odour and various waste management problems were filed. The study population consisting of residents living within 3 km of the landfill boundary was categorised according to the frequency of odour complaints during 1984–1985. The highest odour complaint zone was located within 0.6 miles from the landfill and was consistent with landfill perimeter vinyl chloride measurements, and the local topography and micro-climate.

After adjustment for the effects of parental education, income, race and a number of other potential influencing factors, the average birth weight and gestational age of babies born within the highest odour complaint zone during the entire study period (1978–1986) were similar to those for the reference population characterised by virtually no complaints. The period 1981–1984 was considered the period of highest potential exposures, corresponding to the period of intense disposal activity prior to establishment of effective gas controls including gas abstraction and combustion. The 226 babies born within the highest odour complaint zone and conceived during 1981–1984 had significantly depressed average birth weights and significantly shorter gestational ages compared with the reference population ($n = 8113$). The authors of the study were not able to dismiss incomplete control of potential influencing factors and chance due to small sample sizes as possible explanations of the results. Others have commented that factors other than landfill exposures may explain variation in the frequency of complaints, and this may introduce bias in this surrogate exposure measure.[26] No information was provided on alternative potential environmental pollution sources in the study area.

Conclusions

Rigorous epidemiological studies are extremely difficult to achieve and, by their very nature, can not demonstrate causality.

Failure to account for all landfills within study areas, particularly those within supposedly unexposed reference population areas, leads to difficulties in interpretation relating to the generality of any significant associations. Identifying appropriate reference areas where landfills are absent presents a particular problem in industrialised countries. For example, the SAHSU study indicates that only 20% of the population of England, Scotland and Wales resides more

[26] M. Vrijheid, *Potential Human Health Effects of Landfill Sites, Report to the North West Region of the Environment Agency*, Environmental Epidemiology Unit, London School of Hygiene and Tropical Medicine, London, 1998.

than 2 km away from one form of landfill or another. In some regions the only solution may be to select reference areas on the basis of an absence of the *specific* type of landfill under study, although the lack of a clear distinction between 'hazardous' and household or municipal landfills for old landfill sites may limit this approach for hazardous landfill studies. The SAHSU study is the only study reviewed that appears to have adequately addressed this issue.

The presence of alternative potential industrial sources in areas assumed to be subject to landfill exposures represents an even greater difficulty, which may be unavoidable in single-site studies initiated as a response to community health concerns. In such studies, the only means of identifying the source or cause of apparent associations is to undertake exposure assessments for subsets of the surrounding population. The Nant-y-Gwyddon study included limited exposure measurements in the surrounding community, with inconclusive results. Some off-site environmental pollutant measurements were taken near the Lipari landfill, but concentration levels were not presented or evaluated in the epidemiological study. Multi-site studies would require multivariate statistical methods incorporating indicators of exposures to alternative sources as additional independent variables. None of the reviewed studies adequately addressed this issue, and consequently none were able to satisfactorily distinguish between landfill and other potential industrial sources as the likely source of the purported excess risks.

The SAHSU health data, in particular, appear to be subject to potential confounding effects or other statistical artefacts relating to rare, highly variable health outcomes. For certain outcomes, these difficulties manifested themselves in the appearance of apparently depressed risks in the vicinity of landfills or results that were dependent on whether potential confounding factors were included in the statistical models.

Variation amongst the studies in respect of the criterion chosen as the surrogate for exposure presents further interpretational difficulties. In contrast to the Nant-y-Gwyddon and EUROHAZCON studies, which defined the exposure zone as a 3 km radius, the SAHSU study used a 2 km radius on the basis that this distance was assumed to represent the likely limit of dispersion for landfill emissions. If this assumption is correct, then Vrijheid *et al.*'s data on chromosomal anomalies in the 2–3 km zone, which showed higher risks than more proximal data, cannot be explained by landfill operations. The Lipari and BKK landfill studies used exposure zones confined to less than 1 km from the perimeter of the respective landfill.

For these reasons, it is suggested that further epidemiological studies using more rigorous approaches for defining appropriate reference areas, accounting for alternative industrial sources, incorporating appropriate sensitivity analyses and defining landfill-related exposures are required. In the meantime, the conclusion must be that the studies reviewed do not provide entirely convincing, rigorous epidemiological evidence for an association between landfill and adverse birth outcomes. The authors of the SAHSU study, themselves, suggested that further understanding of the toxicity of landfill emissions and exposure pathways analyses are required to differentiate between the various alternative explanations. Such approaches are discussed in the following section.

4   Theoretical Basis of Purported Effects

This section addresses the question, are sources of emissions sufficient to lead theoretically to public exposures exceeding health criteria? This question is approached using an exposure pathways analysis framework which recognises that public health impacts can only arise at a site if there is a complete exposure pathway that links a source of one or more harmful pollutants at a site with a human receptor who inhales, ingests or dermally absorbs the pollutant. The initial question becomes, are there any complete exposure pathways that are of sufficient magnitude to give rise to adverse health effects in off-site receptors residing in the vicinity of landfill sites?

Potential Exposure Pathways at Landfills

The exposure pathway approach used in this section is based on the US EPA's general risk assessment framework for contaminated sites, as used in the Superfund program.[27] As part of that framework methodology, one considers all the potential exposure pathways that may occur at the type of site under investigation, and then uses knowledge about the circumstances at the site and professional judgement to ascertain which potential exposure pathways would be complete at the site and would therefore require further investigation. Applying that framework to the particular circumstances at landfills, Figure 1 and Table 6 summarise the primary potential exposure pathways that may occur as a result of landfill operations. Table 6 (p. 132) shows the likelihood that each pathway would be complete at controlled landfills in general, and the special circumstances that would lead to a greater likelihood for the complete pathway to exist at a specific landfill site.

Table 6 indicates that inhalation of atmospheric landfill gas emissions is the only exposure pathway that is likely to be common to the majority of landfills. Whilst other pathways may potentially be complete under some circumstances, they would not be a pervasive feature of landfills in general due to the special conditions that they require (as well as, in the case of modern controlled landfills, the presence of standard controls that either eliminate or substantially reduce the magnitude of the pathway).

Moreover, if one accepts the epidemiological evidence for a link between landfill and adverse health effects, then the landfill gas emissions pathway is the most consistent with that evidence. This pathway was the primary exposure pathway that drove the majority of the single-site epidemiological studies (see Table 2a). Consequently, the study areas assumed to represent exposed receptor populations in both the single- and multi-site studies occurred up to 4 km from a landfill, with an average distance of 1.8 km (see Tables 2a and 2b). Pathways involving airborne dust migration, subsurface gas migration, and direct contact would not be expected to occur over such distances, and exposures to water resources would be related less to distance from the site than to the particular

---

[27] US Environmental Protection Agency (US EPA), *Risk Assessment Guidance for Superfund: Volume I, Human Health Evaluation Manual (Part A), Interim Final*, US EPA, Office of Emergency and Remedial Response, Washington DC, 1989.

Figure 1 Potential exposure pathways at landfill sites

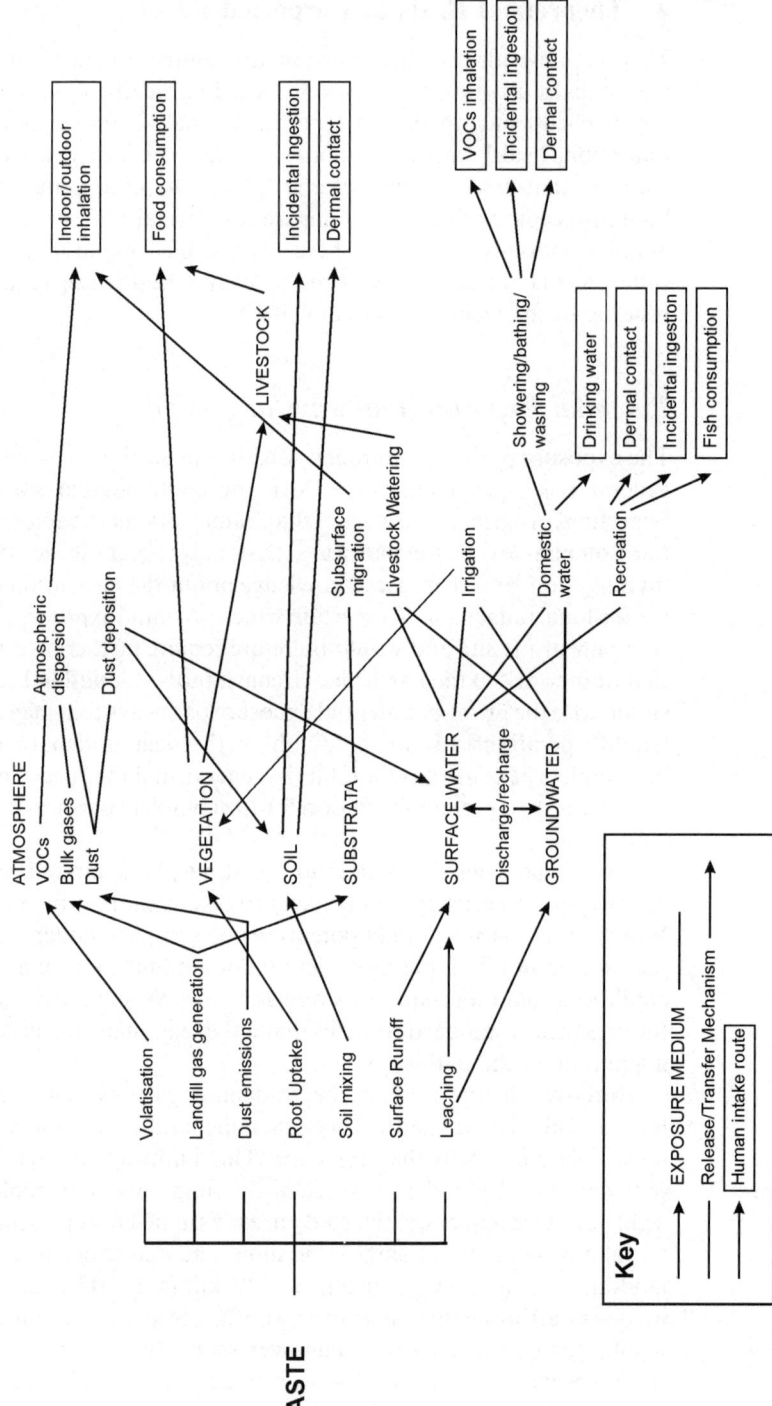

usage catchment area of the affected water abstraction location or distance to the affected surface water course.

The remaining discussion in this section thus focuses on inhalation exposures to aerial emissions of landfill gas and its potentially harmful constituents.

Risk Assessments for Landfill Gas Emissions

Whilst landfill gas consists primarily of methane (CH_4), and carbon dioxide (CO_2), it also contains a wide range of other organic gases in trace amounts, some of which may cause adverse health effects (Table 7, p.134). It is these trace gas constituents that are of concern in public health assessments rather than the bulk gases.

Exposure to trace landfill gas constituents is dependent upon the atmospheric dispersion of the gaseous emissions between the emission source and the exposure point or receptor location (*e.g.*, residential areas, schools, hospitals). The landfill gas and its trace constituents become diluted as the gas mixes and disperses in the atmosphere. The degree of dilution or, conversely, the level of exposure is a function of the rate of gaseous emissions into the atmosphere, distance and orientation between the source of the gas and the receptor location, and also on climatic conditions (wind speed and direction and atmospheric stability).

Scott *et al.* estimated the atmospheric dilution required to reduce raw trace gas concentrations measured at three UK municipal landfill sites to below relevant toxicity thresholds for inhalation exposures, and concluded that the dilution required would be 430-fold for the most toxic compound observed, methanethiol, and that such dilutions would usually be available above landfill sites.[28] Similarly, Young and Parker analysed landfill gas trace constituent data for 6 UK landfills and concluded that a 100-fold dilution would be required to eliminate long-term health hazards for sites receiving industrial wastes.[29] Cumulative effects of multiple trace gas constituents, particularly synergistic effects, where the combined toxicity is greater than the sum of the toxicities of the individual chemicals, would increase the required dilutions indicated by these studies. In addition, the toxicity thresholds used by these studies were US or UK occupational standards, which for public health assessment purposes would need to be adjusted to account for long-term, continuous rather than work-day exposures and for sensitive individuals within the population rather than healthy workers. However, reported average available dilution factors based on methane measurements taken at 1–2 m above the surface of landfills exceed the indicated required dilutions by at least two orders of magnitude and significant further dilution would occur as the landfill gas is dispersed to off-site areas.[28,30]

Crouch *et al.* used atmospheric modelling and exposure assessment approaches to estimate cancer risks to off-site individuals resulting from inhalation of trace

[28] P. E. Scott, C. G. Dent and G. Baldwin, *The Composition and Environmental Impact of Household Waste Derived Landfill Gas: Second Report*, Report No. CWM 041/88, Department of the Environment and Environment Agency, 1988.

[29] P. J. Young and A. Parker, *Waste Manage. Res.*, 1983, **1**, 213.

[30] P. J. Young and L. A. Heasman, in *Proceedings of the GRCDA 8th International Symposium on Landfill Gas*, San Antonio, Texas, 1985.

Table 6 Potential exposure pathways at landfill sites

Exposure pathway	Source	Release mechanism	Transport mechanism	Potential receptors at risk	Exposure route	Likelihood of complete pathway	Circumstances leading to potential complete pathway
Aerial gaseous emissions	• Decomposing waste	Emission of trace constituents in landfill gas	Atmospheric dispersion	Residences, schools, hospitals, OAP homes within up to approx. 3 km of landfill	Inhalation	Moderate: • Containment and treatment controls can reduce but cannot eliminate aerial gaseous emissions	• Sensitive receptors located in vicinity of site
Subsurface gas migration	• Decomposing waste	Emission of bulk and trace constituents in landfill gas	Subsurface migration	Properties within up to approx. 500 m of landfill	Inhalation	Low: • Migration generally limited by natural barriers (e.g. substrata, surface water courses, groundwater table) • Emissions limited by standard controls	• Sensitive receptors located immediately adjacent to site • Site underlain by extensively fractured/fissured strata • Site linked to receptors by man-made structures (sewers, drains etc.)
Airborne dust	• Unpaved haul roads • Soil stockpiles • Bare earth • Earthworks • Dusty waste inputs	Disturbance, by wind or mechanically, of surface dust onto which contaminants are adsorbed	Atmospheric dispersion	Residences, schools, hospitals, OAP homes, food outlets within approx. 250 m of landfill	Inhalation	Low: • Concentrations diminish rapidly with distance • Emissions limited by standard controls	• Sensitive receptors located immediately adjacent to site • Site receives large inputs of highly dusty, hazardous wastes (e.g. incinerator ash, asbestos, industrial powders)
Deposited dust	See above	See above	See above	See above	• Dermal contact • Incidental ingestion	Low: • See above	See above
Direct contact	• Uncovered waste • Contaminated soils	None required	None required	Onsite intruders	• Dermal contact • Incidental ingestion	Low: • Most sites not accessible by public	• Sites adjacent to residences or schools • Sites with no perimeter fencing or other security measures

Food chain	• Contaminated soils • Uncovered waste • Decomposing waste • Contaminated surface water courses	• Direct uptake into food chain • Deposition of dust	Uptake of contaminants by crops, livestock or fish (may require atmospheric dispersion)	• Destination of produce, meat or fish (households, restaurants *etc.*) • Residents with vegetable gardens, orchards • Anglers consuming fish	• Ingestion of food items	Low: • Highly bioaccumulating toxic contaminants not a feature of majority of landfilled wastes • See dust and surface water	• On-site restored areas managed through grazing • Sites adjacent to agricultural areas • Sites receiving wastes containing toxic, bioaccumulating, persistent contaminants
Pathogens	• Uncovered organic waste • Leachate storage/treatment ponds	• Releases of aerosols, dusts • Contact with flies, birds, rodents *etc.*	• Atmospheric dispersion • Movements of flies, birds, rodents *etc.*	Residences, schools, hospitals, OAP homes primarily within immediate vicinity of site, but also within dispersion zone of biotic vectors	• Inhalation • Contact with biotic vectors or their faeces	Low: • Levels of airborne pathogens diminish rapidly with distance from site • Likelihood of exposures via contact with biotic vectors low • Significant releases effectively limited by standard controls (*e.g.* rapid burial of organic wastes, daily cover)	• Highly sensitive receptors located immediately adjacent to site • Sites receiving large quantities of decomposing waste • Inadequate controls leading to uncovered decomposing waste
Groundwater	Leachate	Migration of leachate to aquifer	Abstraction for household or agricultural (irrigation, livestock watering, fisheries) uses	Catchment area for water supply	• Ingestion of drinking water • Dermal contact • Inhalation of volatile organics • Ingestion of irrigated/watered crops/livestock	Low: • Only complete if contaminants migrate to groundwater resource • Potential generally effectively eliminated by containment controls accompanied by monitoring	• Sites overlying or adjacent to major or minor aquifers • Superficial strata of high leaching potential • Sites in close vicinity of abstractions for household or agricultural uses
Surface water	• Leachate • Earthworks • Unvegetated restored areas	• Discharge of leachate to surface water course • Surface runoff	Abstraction for household or agricultural (irrigation, livestock watering, fisheries) uses	• Catchment area for water supply • Stretches of surface water course used for recreational uses	• Ingestion of drinking water • Dermal contact • Inhalation of volatile organics • Ingestion of irrigated/watered crops/livestock	Low: • Generally effectively eliminated by containment controls accompanied by monitoring	• Sites in close vicinity of abstractions for household or agricultural uses • Superficial strata of high leaching potential • Sites adjacent to major fisheries or recreational surface water courses

Table 7 Typical trace constituents in landfill gas

Chemical groups	Typical examples
Alkanes	Octane, nonane, decane
Alkenes	Nonene, decene, butadiene
Cycloalkanes	Cyclohexane, methylcyclohexane
Cycloalkenes	Limonene, other terpenes
Aromatic hydrocarbons	Benzene, toluene, ethylbenzene, xylene
Halogenated compounds	Dichloromethane, trichloroethylene, dichlorobenzene
Alcohols	Propanol, butanol, methylpropanol
Esters	Ethyl butanoate, methyl pentanoate
Ethers	Methyl ethyl ether, diethyl ether
Carboxylic acids	Ethanoic acid, butanoic acid
Amines	Ethylamine, propylamine
Organosulfur compounds	Carbon disulfide, dimethyl sulfide. methanethiol, hydrogen sulfide
Other oxygenated compounds	Acetone, butanone, methylfuran

gas constituents emitted from eight co-disposal and municipal landfill sites in the USA.[31] Total cumulative lifetime cancer risks from 11 of the known or suspect carcinogenic chemicals occurring at the highest concentrations in the landfill gas samples were estimated for a 'worst-case neighbour' living 100 m from the edge of the waste disposal area. Risks resulting from continuous residence adjacent to the worst of the eight landfill sites over a period of 70 years were estimated to be 2×10^{-5}, or 20 in a million, primarily due to inhalation of vinyl chloride. Inspite of the conservative assumptions used, estimated risks were within the US EPA's generally acceptable excess upper-bound lifetime cancer risk range of between one in ten thousand and one in one million for Superfund applications.[32]

We know of no similar published risk assessment studies that specifically examine birth outcomes, as opposed to cancer. This is primarily because the potential developmental effects of landfill gas constituents are poorly understood, although a review of the teratogenicity of landfill emissions commissioned by the UK Government has recently been published.[33]

Eduljee has used a broadly similar risk assessment approach that incorporates a range of public health criteria, relating to both non-cancer and cancer effects.[34] A risk assessment methodology was developed for examining the effects of household landfill emissions *via* a number of potential exposure pathways,

---

[31] E. A. C. Crouch, L. C. Green and S. G. Zemba, in *Proceedings of the GRCDA 13th International Symposium on Landfill Gas*, Lincolnshire, Illinois, 1990.

[32] US Environmental Protection Agency (US EPA), *Risk Assessment Guidance for Superfund: Volume I, Human Health Evaluation Manual (Part D, Standardized Planning, Reporting, and Review of Superfund Risk Assessments)*, Interim Final, Report No. EPA/540/1-89/002, US EPA, Office of Emergency and Remedial Response, Washington DC, 1998.

[33] F. M. Sullivan, S. M. Barlow and P. R. McElhatton, *A Review of the Potential Teratogenicity of Substances Emanating From Landfill Sites*, commissioned by the Department of Health under the Joint Research Programme on the Possible Health Effects of Landfill Sites, 2001.

[34] G. Eduljee, in *Risk Assessment and Risk Management*, ed. R. E. Hester and R. M. Harrison, *Issues in Environmental Science and Technology*, No. 9, Royal Society of Chemistry, Cambridge, 1999, p. 113.

including inhalation exposures to landfill gas, combustion emissions and dusts, as well as groundwater, surface water and other pathways. As an illustrative case example drawn from landfill siting and operational conditions in the UK, risk assessment results were presented for a household waste landfill with an operating gas abstraction and combustion system. Total cumulative exposures across all selected indicator chemicals (*i.e.* those representing the majority of the risks) and all exposure pathways for off-site residences located 50 m and 1 km from the site boundary were well below toxicological criteria. Off-site exposures for inhalation pathways alone were several orders of magnitude below health criteria. (Estimated overall risks were dominated by the groundwater/ingestion pathway, although it was suggested that this pathway represents an extreme exposure route that does not in fact represent a potential threat over timescales of concern due to the required leachate travel times.)

We have compiled landfill gas data from published and unpublished primary literature sources in order to derive average concentrations of trace gas constituents from 25 landfill sites receiving a range of waste types.[28-30,35-42] The concentrations represent uncontrolled or raw landfill gas emissions and were derived from samples of landfill gas in a concentrated state, *e.g.* from within the waste mass itself or in biogas collection systems. The studies report data on approximately 355 trace gas constituents at the 25 landfill sites.

For the trace gas constituents detected by the above studies, we have also compiled the following available health-based standards and guidelines for inhalation exposures developed by the UK Environment Agency (EA), US EPA and World Health Organisation (WHO):[43-46]

- EA's long-term Environmental Assessment Levels (EALs) for air, derived from

[35] P. Young and A. Parker, in *Hazardous and Industrial Waste Management and Testing: Third Symposium*, American Society for Testing and Materials, Philadelphia, 1984.

[36] P. Scott, L. Cowan, C. MacDonald, S. Lanford and S. Davies, *Research and Development of Landfill Gas Abstraction and Utilisation Equipment – Condensate Analysis and Equipment Corrosion: Volume 3, Site Descriptions and Results*, ETSU Report No. ETSU/B/LF/00150/REP/3, Energy Technology Support Unit for the Department of Trade and Industry, 1995.

[37] ENTEC, *Investigations into Odour Problems at Trecatti Landfill, South East Wales: Final Report*, for Environment Agency, Cardiff, 1998.

[38] B. Eklund, E. P. Anderson, B. L. Walker and D. B. Burrows, *Environ. Sci. Technol.*, 1998, **32**, 2233.

[39] M. R. Allen, A. Braithwaite, and C. C. Hills, *Int. J. Environ. Anal. Chem.*, 1996, **62**, 43.

[40] G. Baldwin and P. E. Scott, in *Proceedings of the 3rd International Landfill Symposium*, CISA, Cagliari, Italy, 1991.

[41] B. I. Brookes and P. J. Young, *Talanta*, 1983, **30**, 665.

[42] J. E. Capenter and J. N. Bidwell, *Proceedings of the 17th Biennial Waste Processing Conference*, ASME, 1996.

[43] Environmental Agency, *Best Practicable Environmental Options Assessments for Integrated Pollution Control*, Environment Agency, Technical Guidance Note (Environmental) E1, The Stationery Office, London, 1997.

[44] US Environmental Protection Agency (US EPA), *Health Affects Assessment Summary Tables: FY 1997 Update*, Report No. EPA/540/R-97-036, US EPA, Office of Solid Waste and Emergency Response, Washington DC, 1997.

[45] US Environmental Protection Agency (US EPA), *Integrated Risk Information System (IRIS)*, US EPA, on-line database, 2001.

[46] World Health Organisation (WHO), *Guidelines for Air Quality*, World Health Organisation, Geneva, 2000.

Figure 2 Percentage of landfill gas trace constituents whose mean concentrations exceed Environmental Assessment Levels

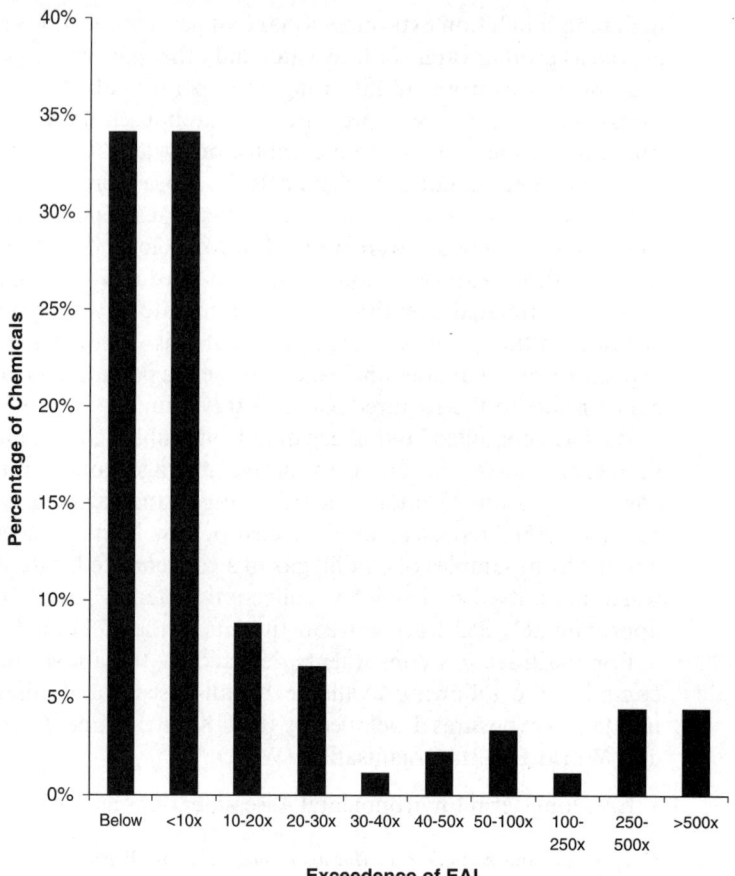

HSE Occupational Exposure Limits (OEL)[47] adjusted to account for (1) long-term exposures to local residential populations compared to those for the workforce and (2) sensitive groups within the general population (*e.g.* children, the elderly and those with diseases such as asthma) (μg m^{-3}).

- US EPA Reference Concentrations for chronic inhalation exposures to non-carcinogenic chemicals; estimate of a daily inhalation exposure of the human population (including sensitive subgroups) that is likely to be without an appreciable risk of deleterious effects during a lifetime (μg m^{-3});
- WHO guideline values for non-carcinogenic effects of contaminants in the air, concentrations below which the risk of occurrence of adverse effects is negligibly low (μg m^{-3});
- US EPA and WHO unit risks of cancer for contaminants in the air; estimates of the probability of an individual developing cancer per unit concentration of a chemical as a result of a lifetime of exposure (risk per μg m^{-3} of air breathed).

The above health criteria are based on toxicological studies on a wide range of

[47] Health & Safety Executive (HSE), *EH40/2000 Occupational Exposure Limits 2000*, HMSO, Norwich, 2000.

health effects including systemic non-cancer effects, cancer and developmental effects. The criteria were derived for public rather than occupational exposures, assuming continuous, chronic exposures and accounting for sensitive sub-groups within the population. Health criteria were available for approximately one-third of the reported trace gas constituents. A proportion of the constituents were detected in raw landfill gas at concentrations exceeding health criteria (Figure 2).

Table 8 provides average raw landfill gas concentrations and health criteria for the ten trace gas constituents that would represent the highest potential risks (determined from the ratio of the average concentration over the most stringent health criteria). The effects on which the various health criteria for the ten chemicals in Table 8 are based include nose and eye irritation (hydrogen sulfide), central nervous system effects (dichloromethane and toluene), liver toxicity (1,1-dichloroethylene and vinyl chloride) and cancer (*e.g.* 1,3-butadiene and benzene). The EAL for trimethylbenzenes is derived from an OEL based on a developmental study that reported reduced foetal body weight following maternal exposures to laboratory animals.[48]

Five of the chemicals in Table 8 (1,3-butadiene, benzene, vinyl chloride, hydrogen sulfide and toluene) have been reported to have demonstrated clear or possible developmental or other reproductive effects in animal and/or human studies.[33] Whilst the criteria values included in Table 8 for these chemicals were not derived from developmental/reproductive studies, they are more stringent than no-observed-effects levels identified from such studies.[33]

Table 8 shows that the dilutions that would be required to reduce average concentrations to below the most stringent health criteria range from 471 for trimethylbenzenes to 4358 for benzene.

Average dilution factors of between approximately 34 000 and over a million between the waste mass and 1.2 m above the surface of uncapped landfills have been reported by one study based on methane measurements.[28] A similar second study reported mean dilutions of between 39 000 and 680 000 available 1–2 m above the landfill surface.[30] These available dilutions are easily sufficient to reduce landfill gas concentrations to below the most stringent health criteria, even immediately above the landfill surface. Atmospheric dilution occurring between the landfill surface and off-site areas would add significant additional margins of safety. Atmospheric dispersion models indicate that dilutions of several thousand-fold are typically available over distances of 50 m, even during unfavourable, stable atmospheric conditions. Still further margins of safety would be provided by landfill gas emissions controls such as capping, active gas abstraction and utilisation/flaring. Methane measurements indicated that capping resulted in a 3.5-fold increase in the dilution between the waste mass and the landfill surface.[28] Active gas abstraction and utilisation/flaring typically result in destruction removal efficiencies for most trace constituents of landfill gas of over 95%.

---

[48] Health & Safety Executive (HSE), *EH64 Summary Criteria for Occupational Exposure Limits*, HMSO, 1996.

Table 8 Comparison of raw landfill gas concentrations with health-based criteria concentrations

Indicator chemical	No. of samples reporting chemical	No. of landfills reporting chemical	Average conc. in raw landfill gas[a] (μg m^{-3})	EAL (μg m^{-3})	Non-carcinogenic criteria (μg m^{-3})		Conc. at 10^{-4} cancer risk[a] (μg m^{-3})		Required dilution for most stringent health criterion (unitless)
					US EPA	WHO	US EPA	WHO	
1,1,2,2-Tetrachloroethane	7	4	5086	Unavail.	Unavail.	Unavail.	1.7	33.3	2992
1,1-Dichloroethylene	35	7	1800	80	Unavail.	Unavail.	2.0	Unavail.	900
1,3-Butadiene	33	7	1705	2.21	Unavail.	Unavail.	0.4	Unavail.	4263
Benzene	61	14	14 119	3.24	Unavail.	Unavail.	12.8	13.3	4358
Chloroethylene (vinyl chloride)	48	12	7446	155	100	Unavail.	11.4	100	653
Dichloromethane (methylene chloride)	41	15	251 763	700	3000	3000	213	Unavail.	1182
Hydrogen sulfide	32	6	3251[b]	140	1	150	Unavail.	Unavail.	3251
Methanethiol	39	13	21 239	10	Unavail.	Unavail.	Unavail.	Unavail.	2124
Toluene	71	22	123 512	1880	400	260	Unavail.	Unavail.	475
Trimethylbenzenes	9	5	578 800	1230	Unavail.	Unavail.	Unavail.	Unavail.	471

[a]Concentrations at 10^{-4} cancer risk were derived by dividing 10^{-4} by the unit risk value.
[b]The average landfill gas concentration for hydrogen sulfide excludes data for two sites with unusually high levels due to the specific type of wastes received.

Conclusion

Whilst there is a range of source–receptor pathways that may potentially result in exposures to pollutant emissions from landfill sites, depending on the special circumstances at a given landfill site, inhalation of atmospheric landfill gas emissions is the only exposure pathway considered likely to be common to the majority of landfills. Inhalation of landfill gas is also the pathway most consistent with epidemiological studies.

The exposure risk assessments described would appear to indicate that emissions of trace gas constituents in landfill gas are not sufficiently high to represent a theoretical basis for adverse health effects in the vicinity of landfill sites (and certainly not at the distances indicated by the epidemiological studies). The dilutions that would occur between the waste mass and the landfill surface and in the atmosphere between the landfill surface and off-site receptors would reduce concentrations below health-based criteria, with margins of safety of several orders of magnitude. Even without engineered controls, the margins of safety appear to be sufficiently wide such that the overall conclusion is unlikely to be affected by the uncertainties associated with the approaches (*e.g.* variation of emissions characteristics within and between landfills, limitations in the toxicological data, synergistic effects, indirect pathways such as food chain uptake of gaseous compounds[49] *etc.*), except under exceptional circumstances.

The exposure risk assessment methodologies described in this section rely on the outputs of atmospheric dispersion models or methane measurements in the air immediately above the landfill surface to predict exposure levels in the ambient environment. Such predictive approaches are inevitably associated with a certain degree of uncertainty. The next logical step is to move from theory towards reality and examine exposures measured in ambient media in communities near landfill sites.

Several authors have stressed the need for systematic measurement of community exposures near landfill sites.[1,2] That view would apply to all potential sources of industrial pollution.[24] However, there are currently few published studies reporting ambient trace gas constituent concentrations in the vicinity of landfills, although a number of studies are on-going. Such studies are required to confirm or refute the predictions from risk assessment models. Of particular importance are epidemiological studies that incorporate direct measures of exposure rather than relying on indirect surrogate measures such as distance, although methodologies for such studies need to be further developed.

5 Summary

It is suggested that confirmation of the existence or magnitude of a causal link between landfill and adverse health effects requires that the evidence satisfies the following criteria: (1) rigorous epidemiological studies, (2) consistency of purported effects, (3) theoretical basis for effects and (4) reality basis for effects.

A number of recent epidemiological studies of UK and European landfill sites

---

[49] A. Eschenroeder and K. von Stackelberg, in *Proceedings of the 84th Annual Meeting of the Air & Waste Management Association*, Vancouver, British Columbia, 1991.

purport elevated risks of certain health effects, including birth defects and low birth weights. However, rigorous epidemiological studies are extremely difficult to achieve and, by their very nature, can not demonstrate causality. It is suggested that further epidemiological studies using more rigorous approaches for defining appropriate reference areas, accounting for alternative industrial sources, incorporating appropriate sensitivity analyses and defining landfill-related exposures are required. In the meantime, the conclusion must be that that the studies reviewed do not provide entirely convincing, rigorous epidemiological evidence for an association between landfill and the health outcomes in question.

Taken as a body of evidence as a whole, the epidemiological data on landfill studies indicate that, on balance, there are a larger number of studies reporting no associations than those reporting positive associations and there is little consistency in terms of individual specific effects. The health outcome coming closest to satisfying the consistency criterion was low birth weight, which is known to be associated with a number of factors that may have a confounding influence on incidence rates in the vicinity of landfills.

Whilst there is a range of source–receptor pathways that may potentially result in exposures to pollutant emissions from landfill sites, inhalation of atmospheric landfill gas emissions is the pathway considered to be most consistent with the epidemiological evidence. Exposure risk assessments for this pathway indicate that predicted ambient air concentrations of trace gas constituents are not sufficiently high to represent a theoretical basis for adverse health effects in the vicinity of landfill sites (and certainly not at the distances indicated by the epidemiological studies). However, more studies examining real ambient concentration measurements in communities near landfill sites are required in order to confirm or refute this conclusion.

Emissions from Solid Waste Management Activities

PAUL T. WILLIAMS

1 Introduction

The management of solid waste represents a major economic and environmental issue throughout the world. Historical trends in waste generation show an increase in the quantities of waste generated for most countries and it is clear that the trend will continue. The treatment and disposal of solid waste involves a range of processes including landfill, incineration, composting, pyrolysis/gasification, *etc.*, all of which may result in emissions to the environment. Even recycling of waste may involve an increase in emissions compared to the production of the virgin product.

This chapter reviews a range of solid waste treatment and disposal processes in terms of the emissions to air, water and land. The processes reviewed are the landfilling of waste and incineration, together with emissions from alternative treatment methods including composting, pyrolysis/gasification techniques, and recycling.

2 Waste Landfill

Landfill represents the largest route for the disposal of waste throughout Europe and North America. Landfill disposal is seen in many respects as the bottom rung of the hierarchy of waste disposal options when considering the concept of sustainable waste management. However, the modern landfill site is an advanced treatment and disposal option designed and managed as an engineering project in which the waste is degraded to a stabilised product and the product leachate is treated to minimise pollution to the environment and the landfill gas is recovered for energy. Older landfill sites, however, were not designed to such standards and may still be emitting pollutants to the environment.

Issues in Environmental Science and Technology, No. 18
Environmental and Health Impact of Solid Waste Management Activities

Decomposition Processes of Municipal Solid Waste in Landfills

Municipal solid waste is a bioreactive type of waste which undergoes biodegradation in the landfill site. Biodegradable materials found in domestic wastes amount to over 65% dry weight and include not only the putrescible food and garden wastes but also paper, cardboard and even to some extent wood and textiles. Industrial and commercial wastes contain up to 62% and 66% of dry weight biodegradable organic material respectively, and consequently would be acceptable in a bioreactive waste landfill site. The processes of degradation of bioreactive waste in landfills takes many years to complete and involves not only biological processes but also inter-related physical and chemical processes.[1-5] Figure 1 shows the major stages of municipal solid waste degradation in landfills.[6]

Stage I. Hydrolysis/aerobic Degradation. The hydrolysis/aerobic degradation stage occurs under aerobic or oxygenated conditions and lasts for only a matter of days or weeks depending on the availability of oxygen for the process which in turn depends on the amount of air trapped in the waste. The micro-organisms are of the aerobic type and they metabolise the available oxygen and a proportion of the organic fraction of the waste to produce simpler hydrocarbons, carbon dioxide, water and heat. The heat generated from the exothermic degradation reaction can raise the temperature of the waste to up to 70–90 °C.[1] Water and carbon dioxide are the main products with carbon dioxide released as gas or absorbed into water to form carbonic acid which gives acidity to the leachate.

Stage II. Hydrolysis and Fermentation. Stage I processes result in a depletion of oxygen in the mass of waste and a change to anaerobic conditions. Different micro-organisms, the facultative anaerobes, which can tolerate reduced oxygen conditions become dominant. Carbohydrates, proteins and lipids, which are the major structural components of bioreactive wastes, are hydrolysed to sugars which are then further decomposed to carbon dioxide, hydrogen, ammonia and organic acids. Proteins decompose *via* deaminisation to form ammonia and also carboxylic acids and carbon dioxide. The derived leachate contains ammoniacal nitrogen in high concentration. The organic acids are mainly acetic acid, but also propionic, butyric, lactic and formic acids and acid derivative products and their formation depends on the composition of the initial waste material. The temperature in the landfill drops to between 30 and 50 °C during this stage. Gas

[1] E. A. McBean, F. A. Rovers and G. J. Farquhar, *Solid Waste Landfill Engineering and Design*, Prentice Hall, New Jersey, 1995.

[2] G. Tchobanoglous and P. R. O'Leary, *Landfilling*, in *Handbook of Solid Waste Management*, F. Kreith (ed.), McGraw-Hill, Inc., New York, 1994.

[3] K. Westlake, *Landfill Pollution and Control*, Albion Publishing, Chichester, 1995.

[4] T. H. Christensen, P. Kjeldsen and B. Lindhardt, *Gas Generating Processes in Landfills*, in *Landfilling of Waste: Biogas*, T. H. Christensen, R. Cossu and R. Stegman (eds.), E. and F. N. Spon, London, 1996.

[5] A. Gendebien, M. Pauwels, M. Constant, M. J. Ledrut-Damanet, E. J. Nyns, H. C. Willumsen, J. Butson, R. Fabry and G. L. Ferrero, Landfill gas: from environment to energy, Commission of the European Communities, Contract No. 88b-7030-11-3-17, Commission of the European Commission, Luxembourg, 1992.

[6] P. T. Williams, *Waste Treatment and Disposal*, John Wiley & Sons Ltd., Chichester, 1998.

Figure 1 Major stages of bioreactive waste degradation in landfills

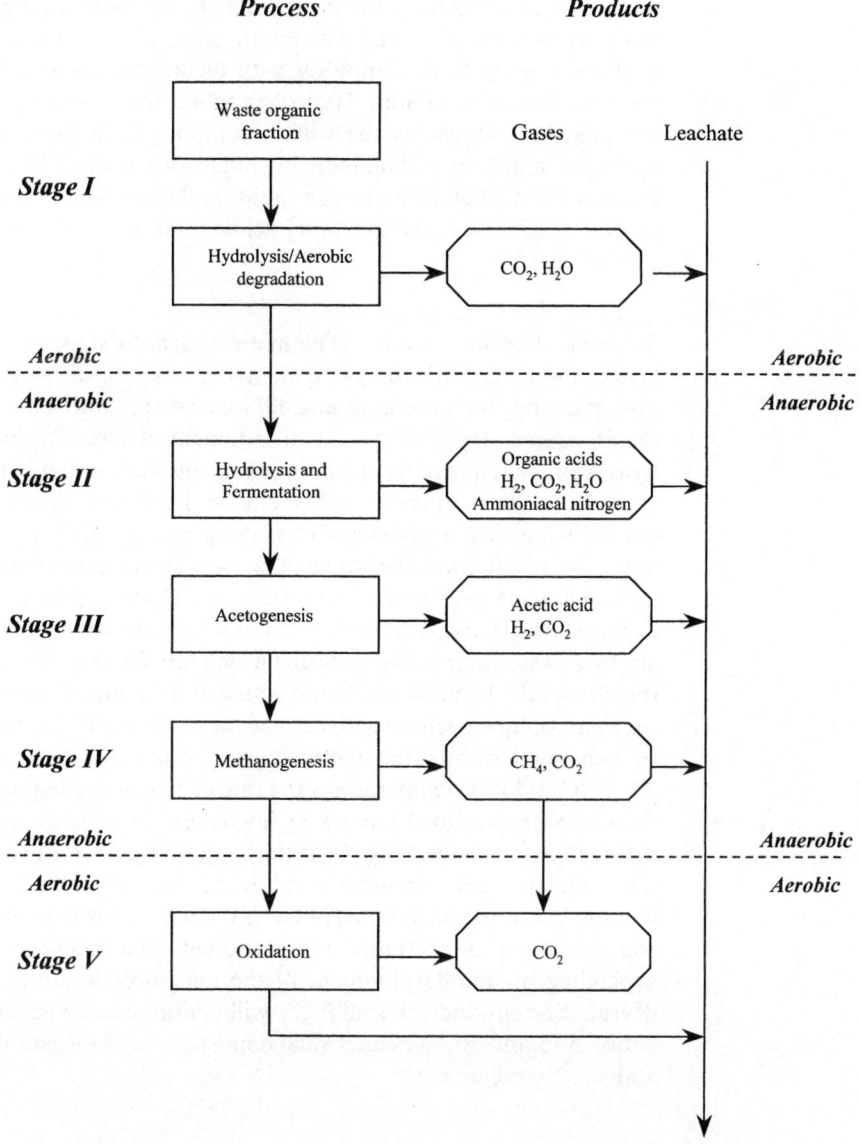

concentrations in the waste undergoing stage II decomposition may rise to levels of up to 80% carbon dioxide and 20% hydrogen.[6]

Stage III. Acetogenesis. The organic acids formed in stage II are converted by acetogen micro-organisms to acetic acid, acetic acid derivatives, carbon dioxide and hydrogen under anaerobic conditions. Other organisms convert carbohydrates directly to acetic acid in the presence of carbon dioxide and hydrogen. Hydrogen and carbon dioxide levels begin to decrease throughout stage III. The acidic conditions of the acetogenic stage increase the solubility of metal ions and increase their concentration in the leachate. In addition, organic acids, chloride ions, ammonium ions and phosphate ions, all in high concentration in the leachate, readily form complexes with metal ions, causing further increases in solubilisation of metal ions. Hydrogen sulfide may also be produced throughout the anaerobic stages as the sulfate compounds in the waste are reduced to hydrogen sulfide by sulfate reducing micro-organisms.[4] Metal sulfides may be a reaction product of the hydrogen sulfide and metal ions in solution. The presence of the organic acids generate a very acidic solution which can have a pH level of 4 or even less.[6]

Stage IV. Methanogenesis.   The methanogenesis stage is the main landfill gas generation stage with the gas composition of typical landfill gas generated at approximately 60% methane and 40% carbon dioxide. The conditions maintain the anaerobic, oxygen depleted environment of stages II and III. Low levels of hydrogen are required to promote organisms, the methanogenic bacteria, which generate carbon dioxide and methane from the organic acids and their derivatives such as acetates and formates generated in the earlier stages. Methane may also result from the direct micro-organism conversion of hydrogen and carbon dioxide to form methane and water. Hydrogen concentrations produced during stages II and III therefore fall to low levels during this fourth stage. There are two classes of micro-organisms which are active in the methanogenic stage, the mesophilic bacteria which are active in the temperature range 30–35 °C and the thermophilic bacteria active in the range 45–65 °C. Therefore landfill gas can be generated during the methanogenic stage over a temperature range of 30–65 °C; at lower temperatures the rate of biological degradation decreases. As the acid concentration becomes depleted the pH rises to about pH 7–8 during the methanogenesis stage. Stage IV is the longest stage of waste degradation but may not commence until six months to several years after the waste is placed in the landfill depending on the level of water content and water circulation. Significant concentrations of methane are generated after between 3 and 12 months depending on the development of the anaerobic micro-organisms and waste degradation products. Landfill gas will continue to be generated for periods of between 15 and 30 years after final deposition of the waste, depending on waste and site characteristics.[6]

Stage V. Oxidation.   The final stage of waste degradation results from the end of the degradation reactions, as the acids are used up in the production of the landfill gas methane and carbon dioxide. New aerobic micro-organisms slowly replace the anaerobic forms and re-establish aerobic conditions. Aerobic micro-organisms which convert residual methane to carbon dioxide and water may become established.

There are a range of factors which influence the degradation of waste in landfill sites and include the depth of the site, the composition of the waste, the degree of prepreparation of the waste such as shredding or compaction, the moisture content of the waste, the movement of the moisture throughout the landfill to distribute the reactants and temperature of the landfill.[3–6]

Table 1 Typical landfill gas composition

Component	Typical value (% by volume)	Observed maximum (% by volume)
Methane	63.8	88.0
Carbon dioxide	33.6	89.3
Oxygen	0.16	20.9
Nitrogen	2.4	87.0
Hydrogen	0.05	21.1
Carbon monoxide	0.001	0.09
Ethane	0.005	0.0139
Ethene	0.018	–
Acetaldehyde	0.005	–
Propane	0.002	0.0171
Butanes	0.003	0.023
Helium	0.00005	–
Higher alkanes	<0.05	0.07
Unsaturated hydrocarbons	0.009	0.048
Halogenated compounds	0.00002	0.032
Hydrogen sulfide	0.00002	35.0
Organosulfur compounds	0.00001	0.028
Alcohols	0.00001	0.127
Others	0.00005	0.023

Source: Waste Management paper 27, 1994.[7]

Landfill Gas

Gases arising from biodegradation landfills consist of mainly hydrogen and carbon dioxide in the early stages followed by mainly methane and carbon dioxide in the later stages; however, a wide range of other gases can be potentially formed and the gas is also saturated with moisture. Table 1 shows the composition of the major constituents of landfill gas.[7] Other minor components identified in landfill gas include, benzene, toluene, xylenes, chloromethane, chlorobenzene, dichlorobenzene *etc.*[6] Minor components such as hydrogen sulfide, organic esters and the organosulfur compounds give landfill gas its characteristic malodorous smell.

Gases generated in the landfill will move throughout the mass of waste in addition to movement or migration out of the site. The mechanism of gas movement is *via* gaseous diffusion and advection or pressure gradient. That is, the gas moves from high to low gas concentration regions or from high to low gas pressure regions. Movement of gas within the mass of waste is governed by the permeability of the waste, overlying daily or intermittent cover and the degree of compaction of the waste. Lateral movement of the gases is caused by overlying low permeability layers such as the daily cover and surface and sub-surface accumulations of water. Vertical movement of gas may occur through natural settlement of the waste, between bales of waste if a baling system is used to compact and bale the waste or through layers of low permeability inert wastes such as construction waste rubble.[7]

[7] Waste Management Paper 27, *Landfill Gas*, Department of the Environment, HMSO, London, 1994.

Gas migration out of the mass of waste into the surrounding environment may occur from older sites or where significant leakage occurs. Migration of gas outside the site requires migration pathways such as high-permeability geological strata, through caves, cavities, cracks in the overlying capping layer, through man-made shafts such as mine shafts and service ducts *etc*. The major components of landfill gas, methane and carbon dioxide, are 'greenhouse gases'. The greenhouse effect is produced by certain gases in the atmosphere which allow transmission of short wave radiation from the sun but are opaque to long wave radiation reflected from the Earth's surface, thereby causing warming of the Earth's atmosphere. A molecule of methane has approximately 30 times the greenhouse effect of a molecule of carbon dioxide.[8] In addition, landfill gas contains components which are flammable and toxic and uncontrolled leakage of landfill gas into houses, shafts, culverts, pipework *etc*. can lead to problems associated with explosive and asphyxiation hazards.

The control of landfill gas to protect the environment and prevent unacceptable risk to human health involves some form of control system. Where the landfill gas is required for an energy recovery system, the landfill gas will be collected *via* migration of the gas to gas pits or wells within the waste which consist of highly permeable gravel, stones or rubble and a gas pump which draws the gas through the gas pit and well system to the surface. The gas is then treated to upgrade the gas through a range of processes depending on the required end-use for the gas. This may involve a condensate removal system, particulate filter, absorption and adsorption systems to scrub the gases and other gas clean-up systems such as membranes and molecular sieves to remove carbon dioxide and trace contaminants.[9,10]

In many cases landfill gas is flared without energy recovery to destroy the methane and organic micro-pollutants as a means of gas hazard and odour control. In addition, the flare may be required to burn off any excess gas or to act as a standby for any plant shutdowns. The flare may be an exposed open flame, usually on a pedestal or enclosed in a ceramic furnace. The open type flare has to maintain a flame even under extremes of weather conditions. The stability of the flame is related to the gas composition, weather conditions, burner design *etc*. However, the flame will be stable at methane concentrations of between 30 and 60%. Minimum flame temperatures of between 850 and 1100°C are recommended to destroy any hazardous trace components.[6]

Where landfill gas concentrations are low such as in older sites or for landfill sites which are used for non-bioreactive wastes such as inert materials, passive venting systems are used. The passive venting pit consists of, for example, a highly permeable vent of gravel material where the gases flow up the vent passively into the atmosphere through a permeable capping layer of sand and granular soil or crushed stone. Construction of the passive venting system may be as emplacement of the waste proceeds or afterwards by drilling or excavation into the mass of waste.

[8] A. Porteus, *Dictionary of Environmental Science and Technology*, John Wiley and Sons Ltd., Chichester, 2000.

[9] K. A. Brown and D. H. Maunder, *Water Sci. Technol.*, 1994, **30**, 143–151.

[10] R. Stegmann, *Landfill gas utilisation: An overview*, in *Landfilling of Waste: Biogas*, T. H. Christensen, R. Cossu and R. Stegmann (eds.), E. and F. N. Spon, London, 1996.

Table 2 Pollutants emissions from the combustion of landfill gas from three UK landfill sites in different types of power plant

Emission	Dual fuel diesel engine[1] (mg m^{-3})	Spark ignition engine[2] (mg m^{-3})	Gas turbine[3] (mg m^{-3})
Particulate matter	4.3	125	9
Carbon monoxide	800	~10 000	14
Unburnt hydrocarbons	22	>200	15
Nitrogen oxides	795	~1170	61
Hydrogen chloride	12	15	38
Sulfur dioxide	51	22	6
Dioxins (ng m^{-3})	0.4	0.6	0.6
Furans (ng m^{-3})	0.4	2.7	1.2

Gas pre-treatment: [1] Drying, filtering, compression and cooling of gas prior to use
[2] Drying, filtering, compression and cooling of gas prior to use
[3] Wet scrubbing, compression, cooling, filtering and heating to 70 °C prior to use.

Source: Blackey, 1993,[11] Young and Blakey, 1996.[12]

Energy recovery from the use of landfill gas via direct use as substitute fuel in boilers, kilns and furnaces and for electricity generation in spark ignition engines, diesel engines or gas turbines involves combustion. The typical composition of landfill gas consists mainly of methane and carbon dioxide with lower concentrations of hydrocarbons. Combustion of the methane in the landfill gas will produce mainly carbon dioxide and water vapour and other minor pollutants. The presence of sulfur, nitrogen and chlorinated organic compounds at trace levels in the landfill gas may lead to the formation of emissions from the combustion of the gas in the energy recovery system. For example Table 2 shows the emissions from the combustion of landfill gas in a dual fuel diesel engine, spark ignition engine and gas turbine.[11,12] The presence of dioxins and furans from the combustion of chlorinated trace components in the landfill gas is significant. The emissions from the combustion of landfill gas will vary depending not only on the type of combustion system, but also the composition of the gas used.

Landfill Leachate

Leachate represents the water which passes through the waste and water generated within the landfill site resulting in a liquid containing suspended solids, soluble components of the waste and products from the degradation of the waste by various micro-organisms. The composition of the leachate will depend on the heterogeneity and composition of the waste and whether there is any

[11] N. C. Blackey, *Emissions from Power Generation Plants Fuelled by Landfill Gas*, in *Sardinia 93*, Fourth International Landfill Symposium (Cagliari, Sardinia, October, 1993), T. H. Christensen, R. Cossu and R. Stegmann (eds.) CISA-Environmental Sanitary Engineering Centre, Cagliari, 1993.
[12] C. P. Young and N. C. Blakey, *Emissions from Landfill Gas Power Generation Plants*, in, *Landfilling of Wastes: Biogas*, T. H. Christensen, R. Cossu and R. Stegmann (eds.), E. and F. N. Spon, 1996.

industrial/hazardous waste co-disposal, the stage of biodegradation reached by the waste, moisture content and operational procedures. Associated with leachate is a malodorous smell due mainly to the presence of organic acids.

The nature of the leachate changes with time as the waste degrades through the various five stages of biodegradation. Table 3 compares the typical leachate of the acetogenic stage III with the methanogenic stage IV.[13] Table 3 shows that the pH of the early formed leachate is acidic/neutral with a pH range between 5.12 and 7.8 equating with the formation of acetic acid and other organic acids by the acetogenic micro-organisms under anaerobic conditions. The organic material of stage III is very high, in the range 1010–29 000 mg l^{-1} for the total organic carbon. Ammoniacal nitrogen levels tend to be higher in stage III due to the biodegradation of the amino acids of proteins and other nitrogenous compounds in the waste. The presence of organic acids of the acetogenic stage increases the solubility of metal ions into the leachate. BOD and COD levels are high with high ratios of BOD:COD indicating that a high proportion of the organic materials in solution are readily biodegradable. Methanogenic leachate has a neutral/alkaline pH reflecting the degradation of the organic acids of stage III to methane and carbon dioxide by the methanogenic micro-organisms. As a consequence, the total organic carbon in the leachate decreases compared to the acetogenic stage. Metal ions continue to be leached from the waste but as the pH of the leachate increases the metal ions become less soluble and decrease in concentration in the leachate. The concentration of ammoniacal nitrogen decreases slightly, but remains high in the leachate. BOD and COD levels decrease compared to acetogenic leachates.

In addition to the components listed in Table 3, a wide range of minor components have been detected in leachate from municipal solid waste. Table 4 shows the concentrations of trace organic compounds found in leachate from a municipal solid waste landfill site.[14,15] Of significance from an environmental viewpoint are the presence of a number of different compounds such as benzene, dioxins and furans of known toxicity.

Landfill leachate has been shown to contain a wide variety of toxic and polluting components. A leachate management system would therefore be required to collect the leachate emanating from the mass of waste before discharge to sewer, water course, land or tidal water. The leachate management system consists of a leachate drainage, collection and treatment system. The drainage of leachate is *via* gravity flow through drainage gradient paths which consist of a permeable granular system containing perforated pipes to collection sumps at low points in the waste mass. The leachate collected in the sumps is then removed by either pumping, gravity drains or side slope risers at the site perimeter. In some cases, the leachate is passed untreated directly to the sewerage facility for off-site treatment. Treatment of leachate to remove the hazardous components can involve a range of processes, including adsorption onto high

[13] Waste Management Paper 26B, *Landfill Design, Construction and Operational Practice*, Department of the Environment, HMSO, London, 1995.

[14] K. Rugge, P. L. Bjerg and T. H. Christensen, *Environ. Sci. Technol.*, 1995, **29**, 1395–1400.

[15] P. R. White, M. Franke and P. Hindle, *Integrated Solid Waste Management; a Lifecycle Inventory*, Blackie Academic and Professional, London, 1995.

Table 3 Composition of acetogenic and methanogenic leachate from large landfill sites with high waste input rate, relatively dry environments (mg l^{-1})

Parameter	Acetogenic			Methanogenic		
	Minimum	Maximum	Mean	Minimum	Maximum	Mean
pH value	5.12	7.8	6.73	6.8	8.2	7.52
COD	2740	152000	36817	622	8000	2307
$BOD_{5\ day}$	2000	68000	18632	97	1770	374
Ammoniacal-N	194	3610	922	283	2040	889
Chloride	659	4670	1805	570	4710	2074
$BOD_{20\ day}$	2000	125000	25108	110	1900	544
Total organic carbon	1010	29000	12217	184	2270	733
Fatty acids (as C)	963	22414	8197	<5	146	18
Alkalinity (as $CaCO_3$)	2720	15870	7251	3000	9130	5376
Conductivity ($\mu S\ cm^{-1}$)	5800	52000	16921	5990	19300	11502
Nitrate-N	<0.2	18.0	1.80	0.2	2.1	0.86
Nitrite-N	0.01	1.4	0.20	<0.01	1.3	0.17
Sulfate (as SO_4)	<5	1560	676	<5	322	67
Phosphate (as P)	0.6	22.6	5.0	0.3	18.4	4.3
Sodium	474	2400	1371	474	3650	1480
Magnesium	25	820	384	40	1580	250
Potassium	350	3100	1143	100	1580	854
Calcium	270	6240	2241	23	501	151
Chromium	0.03	0.3	0.13	<0.03	0.56	0.09
Manganese	1.40	164.0	32.94	0.04	3.59	0.46
Iron	48.3	2300	653.8	1.6	160	27.4
Nickel	<0.03	1.87	0.42	<0.03	0.6	0.17
Copper	0.020	1.10	0.130	<0.02	0.62	0.17
Zinc	0.09	140.0	17.37	0.03	6.7	1.14
Cadmium	<0.01	0.10	0.02	<0.01	0.08	0.015
Lead	<0.04	0.65	0.28	<0.04	1.9	0.20
Arsenic	<0.001	0.148	0.024	<0.001	0.485	0.034
Mercury	<0.0001	0.0015	0.0004	<0.0001	0.0008	0.0002

N.B. Between 13 and 35 samples of acetogenic leachate and 16 and 29 samples of methanogenic leachate analysed to obtain minimum, maximum and mean results.
Units mg l^{-1} except for pH and conductivity.
Source: Waste Management paper 26B, 1995.[13]

surface area ativated carbons, lagoonal evaporation of water to concentrate pollutants, biological filtration, addition of oxidising agents, reed bed biodegradation *etc.* Table 5 shows the main treatment processes for leachate.[6,13,16]

The discussion of the management and treatment of leachate is concerned with

[16] O. Hjelmar, L. M. Johannessen, K. Knox, H. J. Ehrig, J. Flyvbjerg, P. Winther and T. H. Christensen., *Composition and Management of Leachate from Landfills within the EU*, in *Sardinia 95*, Fifth International Landfill Symposium, (Cagliari, Sardinia, October 1995), T. H. Christensen, R. Cossu and R. Stegmann, (Eds.), CISA-Environmental Sanitary Engineering Centre, Cagliari, Sardinia, 1995.

Table 4 Trace organic components found in municipal solid waste leachate

Component	Organic component (mg l^{-1})
1,1,1-Trichloroethane	0.086
1,2-Dichloroethane	0.01
2,4-Dichloroethane	0.13
Benzo[*a*]pyrene	0.00025
Benzene	0.037
Chlorobenzene	0.007
Chloroform	0.029
Chlorophenol	0.00051
Dichloromethane	0.44
Endrin	0.00025
Ethylbenzene	0.058
2-Ethyltoluene	0.005
Hexachlorobenzene	0.0018
Isophorone	0.076
Naphthalene	0.006
Polychlorinatedbiphenyls (PCBs)	0.00073
Pentachlorophenol	0.045
Phenol	0.38
1-Propenylbenzene	0.003
Tetrachloromethane	0.2
Toluene	0.41
Toxaphene	0.001
Trichloroethane	0.043
Vinylchloride	0.04
Xylenes	0.107
Dioxins/furans Toxic Equivalent (TEQ)	0.32 ng

Source: Rugge *et al.*, 1995,[14] White *et al.*, 1995.[15]

containment type landfill sites where a liner system is designed to contain the emissions from the waste degradation processes within the confines of the lined site. In older sites, however, where no lining system was employed then leachate passes directly into the local subsurface environment. In such cases the biological, physical and chemical degradation processes which operate on the leachate as it slowly disperses into the surrounding environment would serve to reduce and dilute the pollution levels to lower levels.[3,6,17]

Biological attenuation processes consist of the continued aerobic and anaerobic biodegradation of the organic components of the leachate as the leachate passes into the surrounding geological and hydrogeological environment. The circulation of groundwater serves to disperse the micro-organisms and nutrient organic material and remove degradation products. Physical attenuation processes result from the dilution of the leachate as it migrates and disperses from the landfill site and becomes diluted by the surrounding groundwater. Dispersion takes place both on a macroscale and microscale. For example, differential dispersion will

[17] A. Bagchi, *Design, Construction and Monitoring of Landfill*, John Wiley and Sons Ltd., New York, 1994.

Table 5 Treatment processes for leachate

Process	Examples
Physico-chemical	
Air stripping of ammonia	Leachate pH adjusted to 11 followed by aeration and release of ammonia gas to a scrubbing unit or atmosphere
Activated carbon adsorption	Highly porous activated carbon adsorbs organic components, used for final stages of treatment
Reverse osmosis	Ultrafiltration membranes concentrate pollutants into a concentrated solution for disposal, used for suspended material, ammoniacal nitrogen, heavy metals
Evaporation	Concentration of contaminants
Oxidation	Addition of oxidising agents such as hydrogen peroxide or sodium hypochlorite solution. Used for sulfides, sulfite, formaldehyde, cyanide and phenolics
Wet air oxidation	Used for high organic content leachates, combustion with air at temperatures up to 310 °C and 200 bar.
Coagulation, flocculation	Addition of reagents followed by mixing and settlement
Attached growth processes	
Trickling filters	Trickling or percolating filters allow the leachate to pass over a substrate containing aerobic micro-organisms which biodegrade the organic components of the leachate
Rotating biological contactors	Rows of rotating discs with attached aerobic micro-organisms
Non-attached growth processes	
Aereation in lagoons	Aerobic micro-organisms in suspension biodegrade the organic constituents in aerated lagoons or tanks
Anaerobic treatment	
Anaerobic biodegradation	Utilises anaerobic methanogenic type micro-organisms to biodegrade the constituents of the leachate
Anaerobic/aerobic treatment	
Reed Bed biodegradation	Reed bed plant systems stimulate the growth of aerobic micro-organisms at the root system and anaerobic micro-organisms in soil areas away from the roots
Land treatment	
Spraying leachate onto land	Spray irrigation of leachate to grassland and woodland, used for low contaminant leachate
Leachate recirculation	
Recirculation of leachate	Recirculation of the leachate through the mass of waste, for further biodegradation

Source: Hjelmar *et al.*, 1995,[16] Waste Management Paper 26B, 1995,[13] Williams, 1998.[6]

151

occur in different types of rock; it will also be influenced by rock grain size, pore size distribution, concentration of clay minerals in the rock *etc.* Leachate may be absorbed or adsorbed into surrounding geological environment and the degree of absorption or adsorption will increase with higher contents of clay minerals or increased concentrations of organic carbon in soils or geological strata. Chemical attenuation processes rely on interaction between leachate and the surrounding geochemical environment to chemically alter or fix the leachate. Interaction of cations and anions in the leachate with those in the soil or rock may occur *via* ion exchange. For example, heavy metals may ion exchange with cations found naturally in soils, clays or different consolidated rock types. Metals may also be removed by precipitation reactions, for example, many metal carbonates, hydroxides and sulfides are insoluble. Chemical reaction between metals in the leachate and such anions in solution results in precipitation. Co-precipitation may also adsorb or occlude trace metals from the leachate within the primary precipitate. Acidic conditions tend to solubilise metals whereas more alkaline conditions induce precipitation. Oxidation–reduction reactions may also occur between inorganic species in the leachate and the surrounding geological and soil environment. Some elements and compounds can exist in more than one oxidation state and changes in the redox potential may influence their mobility into and out of solution.

The potential of biological, physical and chemical processes to reduce the pollution levels of leachate have been demonstrated by Rugge *et al.*[14] They report on the leachate plume emanating from an old municipal solid waste landfill site which penetrated the subsurface environment to a downgradient distance of between 200 and 250 m. The extent of the plume was detected by increased concentrations of inorganic species such as chloride, and more than 15 organic compounds of potential harm to the environment, including benzene, toluene, xylenes, chlorinated hydrocarbons and polycyclic aromatic hydrocarbons, were identified at the subsurface border of the waste and sub-rock. However, 60 m downgradient of the landfill, the plume of hydrocarbons contained negligible concentrations of these hydrocarbons due to the natural dispersion and attenuation processes operating in the subsurface environment.

The implementation of new regulations originating from Europe *via* the European Waste Landfill Directive will influence the emissions from waste landfill sites. The Directive sets out to reduce the landfilling of biodegradable waste and to ensure that the landfill gas produced in existing as well as new landfills is collected, treated and used. The reduction in biodegradable wastes going to landfill has a central aim of reducing emissions of methane as a major greenhouse gas in line with a further European Community Strategy in relation to climate change and reduction of methane emissions. A limit on the disposal of biodegradable waste has been introduced with the aim of encouraging the separate collection of biodegradable waste and to reduce landfilling of waste in general. The Directive requires Member States to reduce landfilling of biodegradable solid waste to 75% of the total produced by 2002, 50% by 2005 and 25% by 2010 from a baseline level taken at 1993. Such a dramatic reduction in biodegradable waste will reduce the overall volumes of landfill gas produced. The requirement for collection, treatment and use of the gas will have a wide impact on the

Table 6 Typical properties of municipal solid waste

Composition	Wt%	Elemental analysis	Wt%
Paper/board	33.0	Carbon	21.5
Plastics	7.0	Hydrogen	3.0
Glass	10.0	Oxygen	16.9
Metals	8.0	Nitrogen	0.5
Food/garden	20.0	Sulfur	0.2
Textiles	4.0	Chlorine	0.4
Other	18.0	Lead (ppm)	101.2
		Cadmium (ppm)	2.4
Proximate analysis		Arsenic (ppm)	5.1
Combustibles	42.1	Zinc (ppm)	225.7
Moisture	31.0	Copper (ppm)	64.0
Ash	26.9	Chromium (ppm)	42.5
		Mercury (ppm)	0.05

Source: Waste Management Paper 28, 1992,[18] Kaiser, 1978,[19] NHWAP, 1994,[20] Buekens and Patrick, 1985.[21]

emissions of landfill gas from waste landfills. Another requirement for waste set out in the Directive is that waste be pre-treated before it is landfilled. Pre-treatment is defined as the physical, chemical or biological processes, including sorting, that change the characteristics of the waste in order to reduce its volume or hazardous nature, facilitate its handling or enhance recovery.

3   Incineration

Mass burn incineration is used for the treatment and disposal of municipal solid waste throughout the world. The composition and characteristics of the waste will influence the combustion properties and emissions produced from the combustion system. A typical calorific value for municipal solid waste is approximately 9000 kJ kg^{-1}; ash and moisture contents tend to be high and thus in terms of a fuel the waste would compare poorly with, for example, coal. Table 6 shows typical properties of municipal solid waste.[18-21] The waste also contains significant concentrations of heavy metals such as cadmium, lead, zinc and chromium, which will influence the emissions of such metals. Similarly, the sulfur and chlorine content will produce emissions of sulfur dioxide and hydrogen chloride.

Figure 2 shows a schematic diagram of a typical mass burn incinerator. The waste is delivered by collection vehicles and tipped into the storage bunker. The travelling type crane loads the waste to the feeding system represented by a steel

[18] Waste Management Paper 28, *Recycling*, Department of the Environment, HMSO, London, 1992.

[19] E. Kaiser, in *Combustion and Incineration Processes*, W. R. Niessen (ed.), Dekker Press, New York, 1978.

[20] NHWAP, 1994 The technical aspects of controlled waste management; National household waste analysis project, Phase 2, Volume 3, Chemical analysis data. Report No. CWM/087/94, Department of the Environment, wastes Technical Division, HMSO, London 1994.

[21] A. Buekens and P. K. Patrick, *Incineration*, in *Solid Waste Management; Selected Topics*, M. J. Suess (ed.), World Health Organisation, Regional Office for Environmental Health Hazards, Copenhagen, Denmark, 1985.

Figure 2 Schematic diagram of a typical mass burn waste incinerator

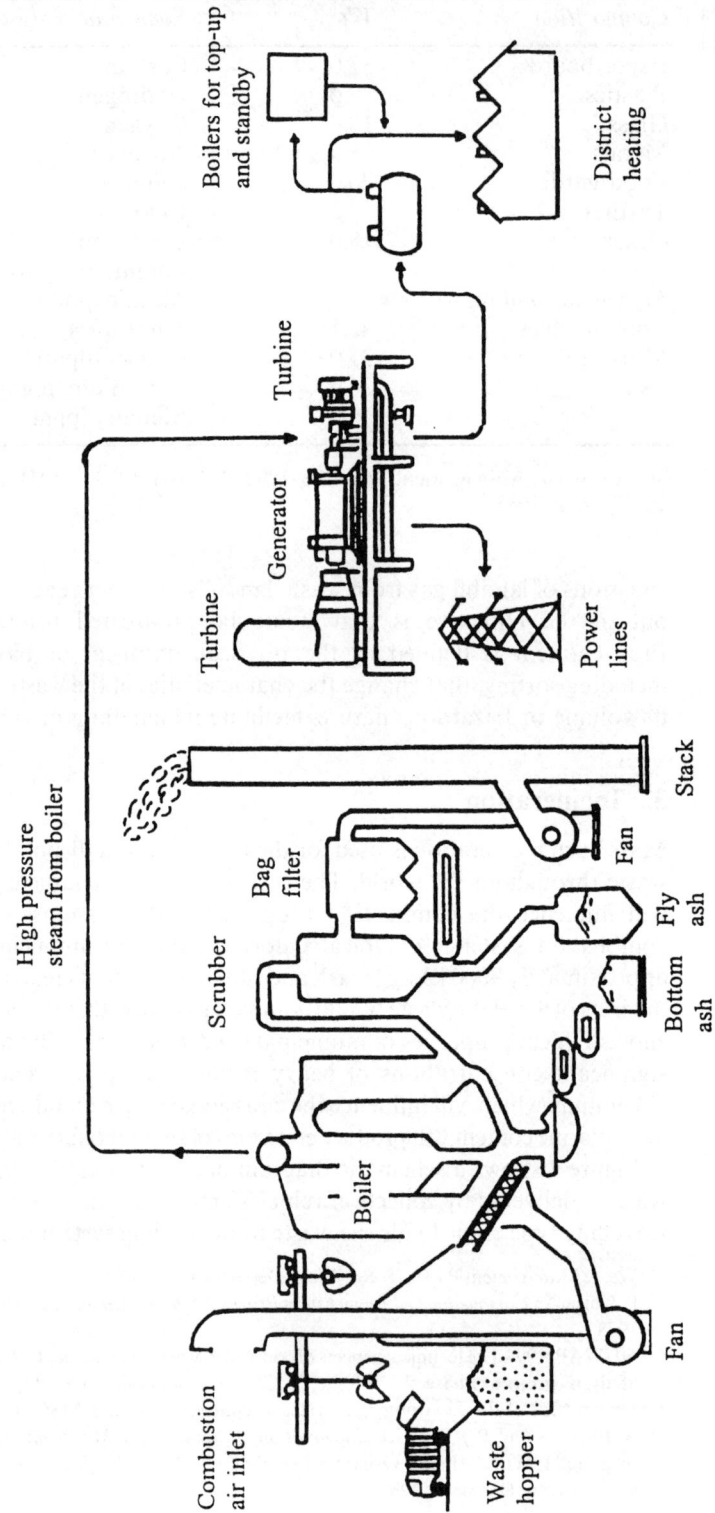

hopper where the waste is allowed to flow into the incinerator under its own weight without bridging or blocking. The waste is fed into the furnace and undergoes three stages of incineration: drying and devolatilisation; combustion of volatiles and soot; combustion of the solid carbonaceous residue. As the waste enters the hot furnace the waste is heated up *via* contact with hot combustion gases, pre-heated air or radiated heat from the incinerator walls and initially moisture is driven off in the temperature range 50–100 °C. After moisture release the waste then undergoes thermal decomposition and pyrolysis of the organic material such as paper, plastics, food waste, textiles *etc.* in the waste, which generates the volatile matter, the combustible gases and vapours. Devolatilisation takes place over a wide range of temperatures from about 200 to 750 °C with the main release of volatiles between 425 and 550 °C. The combustion of volatiles takes place immediately above the surface of the waste on the grate and in the combustion chamber above the grate. Complete combustion of the gases and vapours requires sufficiently high temperature, (typically about 1000 °C), adequate residence time (about 2–4 s) and excess turbulent air to ensure good mixing. After the drying and devolatilisation stages the residue consist of a carbonaceous char which combusts on the grate of the furnace and the inert material. The ash and metals residue is discharged continuously at the end of the last grate section and is known as bottom ash.

The hot combustion gases pass through the boiler system which produces steam and reduces the temperature of the gases. The polluted gases are then cleaned using a variety of gas clean-up measures such as acid gas removal by scrubbing, particulate removal using electrostatic precipitators or fabric (bag) filters. Organic micropollutants such as dioxins and furans and volatile heavy metals such as mercury and cadmium are usually removed by addition of activated carbon onto which the dioxins and furans are adsorbed and the contaminated activated carbon is then trapped by the fabric filter. Removal of nitrogen oxides may require such measure as the addition of ammonia to produce nitrogen and water from the reaction with the nitrogen oxides. The economic viability of incineration as a waste treatment and disposal route for municipal solid waste in most cases depends on the recovery of energy from the process to offset the high costs involved in incineration. Therefore the modern municipal waste incinerator relies on the production of steam for electricity generation or district heating to ensure the cost effectiveness of the process. Some schemes may incorporate both electricity generation and district heating as combined heat and power (CHP) systems.

Of the pollutant emissions arising from the incineration of waste, those emitted to the atmosphere have received most attention from environmentalists and legislators and stringent emission limits are in place to regulate the emissions to air from the incineration of waste. However, waste incineration also produces approximately 30 wt% ash consisting of the bottom ash and the flyash captured in the pollution abatement system. Additionally, wastewater arises from incinerators but only at low levels. The emission limits to air for member states of the European Union are set out in the EU Waste Incineration Directive for the incineration of waste and are shown in Table 7.

The pollution abatement system to control the emissions to air from waste

Table 7 Emission limits for waste incineration as set out in the EU Waste Incineration Directive (reference conditions: 273 K, 101 kPa, 11% oxygen, dry gas)

Pollutant	Emission limit (mg m^{-3})
Total dust	10 (daily average)
Total organic carbon (gaseous and vapours)	10 (daily average)
HCl	10 (daily average)
HF	1 (daily average)
CO	50 (daily average)
SO$_2$	50 (daily average)
NO$_x$ (expressed as NO$_2$) (New plant)	200 (daily average)
NO$_x$ (expressed as NO$_2$) (Existing plant 6–16 th^{-1})	400 (daily average) (until 2010)
NO$_x$ (expressed as NO$_2$) (Existing plant 16–25 t h^{-1})	400 (daily average) (until 2008)
Cd and Tl	0.05 (new plant) 0.1 (existing plant to 2007)
Hg	0.05 (new plant) 0.1 (existing plant to 2007)
Sb, As, Pb, Cr, Co, Cu, Mn, Ni, V (Total)	0.5 (total – new plant) 1.0 (existing plant to 2007)
Dioxins and furans (TEQ) ng m^{-3}	0.1

Source: EU Waste Incineration Directive, 2000.[33]

Table 8 Typical concentration ranges of emissions from municipal solid waste mass burn incineration *before* gas clean-up (mg m^{-3})

Emissions	Minimum	Maximum	Mean
Particulate	1500	8000	3000
Hydrogen chloride	400	2200	1150
Hydrogen fluoride	5	20	9
Sulfur oxides	200	2000	500
Nitrogen oxides	150	650	250
Lead	6	55	30
Cadmium	0.3	3.6	1.8
Mercury	0.1	1.1	0.5

Source: Williams, 1998.[6]

incineration constitutes a major proportion of the cost, technological sophistication and space requirement of an incinerator. The emissions of most concern and which require control to legislated emission limits are total particulate or dust, acidic gases such as hydrogen chloride, hydrogen fluoride and sulfur dioxide, heavy metals such as mercury, cadmium and lead, and dioxins and furans.

Table 8 shows typical concentration ranges for emissions *before* any gas clean-up treatment.[6] The emissions are very much higher than are legally permitted and emphasise the need for efficient and sophisticated gas clean-up to reduce the emissions to below legislated values.

Particulates

The combustion of waste is a very dusty process. The agitation of the waste as it tumbles down the grate, the blowing of primary air through the bed, the high ash content of the waste, and the heterogeneous nature of the waste all serve to produce a high particulate loading in the flue gases. The particulate is largely composed of ash; however, in addition, pollutants of a more toxic nature such as heavy metals and dioxins and furans are associated with particulate matter, adsorbed on the surface of the particle. The particulates may also contain carbon and adsorbed acidic gases such as hydrochloric, sulfuric or even hydrofluoric acid to produce corrosive acid 'smuts'.

The size range of incinerator particulates found in the flue gases is from < 1 μm to 75 μm, the larger particles tending to settle out prior to the flue.[22] It is the ultrafine particles that are of particular concern in assessment of health effects since they contain ash and adsorbed acid gases, heavy metals and organic micropollutants which because of their size can pass deep into the respiratory system of humans. There is currently some concern that the important factor in determining the deleterious effects of fine particles is not particularly their composition but their ultrafine nature and the fact that they can penetrate deep into the lungs. A separate size category of particulate matter of environmental concern, PM_{10}, has been designated for particulate matter of less than 10 μm in size. A large fraction of municipal waste incinerator particulates are of such small size. In addition, their small size promotes both short- and long-range dispersion from the chimney stack into the environment. The most common control solution for particulates in waste incineration is *via* either electrostatic precipitators or fabric filters.

Heavy Metals

Table 6 shows that significant concentrations of heavy metals are present in municipal solid waste. For example, municipal refuse may contain lead from lead based paints, mercury and cadmium from batteries, aluminium foil, lead plumbing, zinc sheets, volatile metal compounds *etc*. The extent of evaporation of these metals and metal compounds in the furnace of a municipal waste incinerator is a function of many factors, including volatility, combustion conditions and ash entrainment.[23] The partitioning of the heavy metals in the incinerator system is a function of their physico-chemical properties. For example, cadmium and mercury, being the more volatile of the heavy metals with high vapour pressures and low boiling points, are most likely to be found in the flue gas.[24,25] Other metals with low vapour pressure and high boiling points such as iron and copper are almost completely trapped in the bottom ash. In addition to the physical properties of volatilisation, simultaneous chemical processes such as decomposition, chlorination, oxidation and reduction may take place.[26,27] For example, Seeker[28] has shown that chlorine influences the volatility of heavy

[22] W. R. Niessen, Combustion and incineration processes., Marcel Dekker Inc., New York 1978.
[23] R. G. Barton, W. D. Clark and W. R. Seeker, *Combustion Sci. Technol.*, 1990, **74**, 327–342.
[24] P. H. Brunner and H. Monch, *Waste Manage. Res.*, 1986, **4**, 105–119.
[25] L. S. Morf, P. H. Brunner and S. Spaun, *Waste Manage. Res.*, 2000, **18**, 4–15.
[26] J. A. Mulholland and A. F. Sarofim, *Environ. Sci. Technol.*, 1991, **25**, 268.
[27] J. A. Mulholland and A. F. Sarofim, *Environ Progr.*, 1991, **10**, 83.

P. T. Williams

metals *via* the formation of chlorides. For example, nickel, because of its low vapour pressure and high boiling point, will not vaporise under the conditions of incinerator furnaces, but will do so in the presence of chlorine.[29] Also, cadmium is easily volatilised during incineration and is oxidised in the presence of hydrogen chloride to form mainly cadmium chloride.[30] Changes in the oxidising and reducing conditions within the incinerator also can influence the volatilisation of heavy metals. Barton *et al.*[23] have suggested that low-volatility metal compounds may react under reducing conditions to form compounds which are more readily volatilised. As the gases cool and conditions change to a more oxygenated environment, conversion back to the original low-volatility compound occurs and condensation results. A further route for the heavy metals to enter the flue gas stream is *via* entrainment of fine ash particles containing cadmium either as the metal or as cadmium compounds. Entrainment is a function of the size, shape and density of the ash particles as well as the incinerator operating conditions. It has also been shown that the matrix of the waste input can have a significant impact on the transfer of metals into the gas phase.[25]

As the furnace off-gases cool as they pass through the flue gas system, the heavy metals are subject to a series of condensation reactions involving homogeneous nucleation to form a fine fume of metal particles and heterogeneous deposition onto flyash.[28] Homogeneous nucleation occurs for example when the partial pressure of an inorganic vapour species exceeds a certain critical value.[31] The incineration gases may become supersaturated as a result of rapid cooling of the gas or rapid formation of a new metal species of lower volatility. Heterogeneous deposition involves flyash particles in the flue gases providing sites for condensation of the cooling metal vapour. It has also been shown that the relative rates of homogeneous nucleation and heterogeneous deposition also depend on the time/temperature gradient experienced by the metal-containing flue gas.[29] Following homogeneous nucleation and heterogeneous deposition, particles will subsequently grow by coagulation.

The gas clean-up systems required to control heavy metals are dependent on the metals' volatility. Measures used to control total particulate emissions such as electrostatic precipitators and fabric filters will collect the associated heavy metals which are in the flyash, adsorbed to the surface or as discrete heavy metal particles. The more volatile heavy metals, particularly mercury and to some extent cadmium, require more sophisticated emissions control equipment. Commonly, activated carbon is added at concentration levels of $0.1–0.5\,g\,m^{-3}$ of waste gas, and high removal efficiencies have been reported.[6] After gas clean-up, such as electrostatic precipitation, activated carbon addition and fabric filtration, heavy metal emissions fall below the legislated limit values as shown in Table 9.[32]

[28] W. R. Seeker, *Waste Combustion*, 23rd Symposium (International) on Combustion, The Combustion Institute, Pittsburgh, 1990, pp. 867–885.
[29] S. B. Davis, T. K. Gale, J. O. L. Wendt and W. P. Linak, 27th Symposium (International) on Combustion, The Combustion Institute, Pittsburgh, 1998, pp. 1785–1791.
[30] H. Vogg, H. Braun, M. Metzger and J. Schneider, *Waste Manage. Res.*, 1986, **4**, 65–74.
[31] T. C. Ho, C. Chen, J. R. Hopper and D. A. Oberacker, *Combust. Sci. Tech.*, 1992, **85**, 101–116.
[32] G. Atkins, Energy from waste as the BPEO; A strategic review of the SELCHP project, SELCHP Waste to energy facility publicity material, London, 1997.

Table 9 Emissions to air from a typical modern UK plant (1997)

Pollutant	Emission (mg m^{-3})
Total particulate matter	1.4
Hydrogen chloride	18.4
Hydrogen fluoride	<0.01
Sulfur dioxide	11.3
Nitrogen oxides	564
Volatile organic compounds	<3
Cadmium	0.001
Mercury	0.006
Total of arsenic, chromium, copper, lead, Manganese, nickel, tin	0.03
Carbon monoxide	9
Dioxins (TEQ)	0.02 (ng m^{-3})

Source: Atkins, 1997.[32]

Toxic and Corrosive Gases

Municipal waste contains a range of compounds which contain chlorine, fluorine, sulfur, nitrogen and other elements which may result in the generation of toxic or corrosive gases such as hydrogen chloride, hydrogen fluoride, sulfur oxides and nitrogen oxides. One of the major sources of HCl is regarded as PVC plastic, together with other sources such as metal chlorides like NaCl or $CaCl_2$ from paper and board, rubber, leather and vegetable matter. PVC emits HCl by a gradual process of thermal decomposition which takes place between 180 and 600 °C. Hydrogen fluoride arises from combustion of fluorinated hydrocarbons such as plastics like PTFE. The sulfur content of municipal solid waste is relatively low and represents only a minor source of sulfur dioxide (SO_2) emission when compared to power plants and industrial boilers firing heavy fuel oil or coal. About 1% of the SO_2 may be further oxidised to sulfur trioxide, SO_3, which reacts with water vapour to form highly corrosive sulfuric acid, H_2SO_4, in the flue gas. Nitrogen oxides arise from the nitrogen in the waste (fuel NO_x) and by direct combination of the nitrogen and oxygen present in the combustion air which occurs more rapidly at high temperature (thermal NO_x). Thermal NO_x is regarded as by far the largest source of NO_x for waste incinerators.

Table 7 shows that the emissions of HCl, HF, SO_2 and NO_x are all regulated and require extensive gas clean-up measures to comply with the legislated emission limits (EU Waste Incineration Directive, 2000[33]). Williams[6] has reviewed clean-up of flue gases including acid gas control from municipal waste incinerators. Wet, dry and semi-dry processes are used to remove the acid gases produced by waste combustion. Wet scrubbing systems use slurries and solutions at relatively low temperatures and produce a liquid or wet solid/sludge reaction product. The adsorbents used include, calcium oxide, calcium hydroxide and

[33] EU Waste Incineration Directive; Directive 2000/76/EC, *Official Journal of the European Communities*, Directive 2000/76/EEC of the European Parliament and of the Council of 4th December 2000 on the Incineration of waste, 28/12/2000, L332/91, 2000.

sodium hydroxide. The liquid or sludge product is highly polluted and difficult and expensive to treat. Consequently, developments have centred on new methods of control which generate a solid residue which is easier to handle. Dry systems use a dry powder and possibly upstream humidification to improve gas/sorbent reaction. Semi-dry processes use an alkaline sorbent slurry or solution which is atomised into fine droplets into the flue gas, the droplets react and dry in the hot flue gases to produce a dry powder. For both the dry and semi-dry systems adsorption is improved by the use of a downstream fabric filter which increases contact time between the gases and the alkaline filter cake formed on the filter by the adsorbent.

Nitric oxide, the main oxide of nitrogen found in flue gases, cannot be reduced by scrubbing because of its low solubility. The addition of ammonia to the combustion gases to form nitrogen and water has been used to reduce NO_x but is only effective in a very narrow temperature range of 870–900°C. At higher temperatures the ammonia itself breaks down to produce NO_x and at lower temperatures the reaction is too slow to be effective. However, flue gas NO_x can be controlled by selective catalytic reduction (SCR) in the presence of added ammonia which reproduces the ammonia reduction reaction but at a lower temperature and wider temperature range. Typical catalysts are platinum, palladium, vanadium oxide and titanium oxide. The ammonia which is added upstream of the catalysts bed reacts with the NO_x in the presence of the catalyst at 300–400°C to produce nitrogen and water. Since the catalyst can be deactivated by heavy metals the de-NO_x system is located after the fabric filter. Selective catalytic reduction can reduce NO_x levels by over 90%.

Dioxins and Furans

Dioxins (polychlorinated dibenzodioxins – PCDD) and furans (polychlorinated dibenzofurans – PCDF) are organic micropollutants which are known to occur at trace levels in waste incinerator flue gases. There is considerable concern from the public and environmental groups surrounding the issue of dioxins and furans in the environment. Recent reviews of PCDD/F releases to the environment[34,35] have identified municipal solid waste incineration as a significant source of dioxin and furan emissions to the atmosphere.

Several mechanisms have been proposed for the formation of PCDD/F in waste incineration systems. For example: (a) high temperature gas phase formation *via* pyrosynthesis; (b) low temperature *de novo* formation from macromolecular carbon and organic or inorganic chlorine present in the flyash matrix; (c) formation from different organic precursors such as chlorophenols, chlorinated benzenes and polychlorinated biphenyls which may be formed in the gas phase during incomplete combustion and which then combine heterogeneously and catalytically with the flyash surface.[36] However, it is thought that the dominant formation route for PCDD/F in municipal solid waste incinerators is 'de novo' synthesis, the reaction occurring in the down stream boiler and pollution

[34] P. H. Dyke *et al.*, *Sci. Total Environ.*, 1997, **207**, 119.
[35] G. Eduljee and P. H. Dyke, *Sci. Total Environ.*, 1996, **177**, 303.
[36] K. Tuppurainen *et al.*, *Chemosphere*, 1998, **36**, 1493.

abatement system. Reaction between HCl or chlorine and an organic substrate catalysed on the surface of flyash particles in an optimised temperature window of between 250 and 400 °C produces PCDD/F.[37] It is thought the reaction is catalysed by various metals, metal oxides, silicates *etc.*, particularly copper chloride present in the fly ash.[6]

Table 7 shows that dioxins and furans are controlled by emission limit values which requires monitoring of the incinerator stack gases to levels of less than 0.1 ng m^{-3} and are reported in terms of their Toxic Equivalent (TEQ). The assessment of the toxicity of PCDD and PCDF mixtures has led to the development of the Toxic Equivalent scheme. This uses the available toxicological and biological data to generate a set of weighting factors each of which expresses the toxicity of a particular PCDD or PCDF in terms of an equivalent amount of the most toxic and most analysed PCDD – 2,3,7,8-TCDD.

Operational and process control are used to minimise emissions of PCDD/F from waste incineration, including combustion control to minimise the formation of PCDD/F precursors and the rapid cooling of the flue gases to below the optimum *de novo* reaction temperature. Activated carbon injection is the preferred post-combustion technology for reducing emissions of PCDD/F from municipal solid waste incinerator flue gases. The activated carbon is used with wet, semi-dry and dry scrubbers for PCDD/F removal; after adsorption on activated carbon, the activated carbon, with flyash, is then captured by fabric filter and with the added benefit of adsorption on the filter cake on the fabric filter surface.[38] Novel methods of dioxin removal incorporating catalytic reaction[39] and catalytically coated fabric filters have also been commercialised.[40] Table 9 shows that with the use of efficient combustion control, flue gas cooling and the use of additive activated carbon, the emission levels of dioxins and furans from municipal solid waste incineration can be reduced to extremely low levels.

Incinerator Wastewater

Relatively lower levels of contaminated wastewater are generated from the waste incineration process compared to emissions to air. The main sources of wastewater from incinerators are from flue gas treatment as flue gas scrubber water and alkaline scrubbing of the gases to remove acid gases and the quenching of incinerator bottom ash. Other minor sources include, for example, scrubber water pre-treatment and the purification of boiler feedwater where a boiler plant is installed. Where the gases are scrubbed or cooled with water the absorbed acid gases will make the water very acidic and will also consequently contain significant quantities of heavy metals which are soluble in the acidic solution. Where the gases are scrubbed with an alkaline solution such as sodium hydroxide or calcium hydroxide to remove acid gases, the scrubber water will be very

[37] G. Eduljee, *Environ. Waste Manage.*, 1999, **2**, 45.

[38] M. B. Chang and J.J. Lin, *Chemosphere*, 2001, **45**, 1151.

[39] M. Goemans *et al.*, *3rd International Symposium on Incineration and Flue Gas Treatment Technologies*, Brussels, 2–4 July, 2001.

[40] G. G. Pranghofer and K. J. Fritsky, *3rd International Symposium on Incineration and Flue Gas Treatment Technologies*, Brussels, 2–4 July, 2001.

Table 10 Emission limit values for the discharges of waste water from the cleaning of exhaust gases

Polluting substance	Emission limit (mg l^{-1})	Wastewater 1st Stage	Wastewater 2nd Stage	Treated wastewater both stages
Total solids	30[1]	–	–	–
Total solids	45[2]	–	–	–
Hg	0.03	0.051	0.02	0.03–0.27
Cd	0.05	<0.004	<0.004	–
Tl	0.05	–	–	–
As	0.15	–	–	–
Pb	0.2	2.6	0.46	<0.10–0.21
Cr	0.5	3.2	0.74	0.10–0.47
Cu	0.5	3.0	0.79	0.002–0.015
Ni	0.5	34	2.0	<0.02–0.68
Zn	1.5	–	–	<0.01–0.11
Dioxins (TEQ)	0.3	–	–	–

[1] 95% values do not exceed
[2] 100% values do not exceed
Source: EU Waste Incineration Directive 2000,[33] IAWG, 1997.[44]

alkaline. In some cases the bottom ash from the incinerator grate is removed in a unit which serves to cool the ash and also maintain a partial vacuum in the incinerator chamber. Bottom ash waste water is alkaline and contains only low levels of dissolved heavy metals, below permitted sewerage discharge levels.[6]

The main pollutants in incinerator waste waters are heavy metals. Whilst there is most concern over the presence of heavy metals in wastewater, the presence of organic pollutants such as dioxins and furans has also been reported. However, levels of PCDD and PCDF are very low as dissolved compounds in waste water,[41,42] but have been detected in significant concentrations in suspended particles, which therefore are required to be filtered out.[42,43]

The European Union Waste Incineration Directive, in addition to limits to air, also specifies emission limit values for discharges of waste water from the cleaning of exhaust gases and are shown in Table 10.[33,44] Also shown in Table 10 are typical compositions of wastewater from the first and second stages of a two-stage wet scrubber system for the clean-up of flue gases and after treatment in a wastewater treatment plant for a 4×12 tonnes h^{-1} municipal solid waste incinerator plant.[44] Clean-up of the heavy metals is usually through neutralisation and precipitation and the resultant precipitate is typically mixed with flyash and other air pollution control residues and landfilled.[41,44]

[41] D. O. Reimann, *Waste Manage. Res.*, 1987, **5**, 147.

[42] V. Ozvacic, G. Wong, H. Tosine, R. E. Clement and J. Osborne, *J. Air Pollut. Control Assoc.*, 1985, **35**, 849.

[43] R. R. Bumb, W. B. Crummett, S. S. Cutie, J. R. Gledhill, R. H. Hummel, R. O. Kagel, L. L. Lamparski, E. V. Luomoa, D. L. Miller, L. A. Nestrick, L. A. Shadoff, R. H. Stehl and J. S. Woods, *Science*, 1980, **210**, 385.

[44] IAWG, International Ash Working Group, *Municipal Solid Waste Incinerator Residues*, International Ash Working Group, Elsevier Publications, London, 1997.

Table 11 Typical composition of bottom ash, flyash and APC residues from a dry/semi-dry system and a wet control system

Polluting substance	Bottom ash	Flyash	APC residue dry/semi-dry	APC residue wet
Aluminium (g kg^{-1})	22–73	49–90	12–83	21–39
Silicon (g kg^{-1})	91–310	95–210	36–120	78
Calcium (g kg^{-1})	37–120	74–130	110–350	87–200
Potassium (g kg^{-1})	0.7–16	22–62	5.9–40	0.8–8.6
Magnesium (g kg^{-1})	0.4–26	11–19	5.1–14	19–170
Sodium (g kg^{-1})	2.9–42	15–57	7.6–29	0.7–3.4
Iron (g kg^{-1})	4.1–150	12–44	2.6–71	20–97
Chlorine (g kg^{-1})	0.8–4.2	29–210	62–380	17–51
Cadmium (mg kg^{-1})	0.3–71	50–450	140–300	150–1400
Chromium (mg kg^{-1})	23–3200	140–1100	73–570	80–560
Copper (mg kg^{-1})	190–8200	600–3200	16–1700	440–2400
Mercury (mg kg^{-1})	0.02–7.8	0.7–30	0.1–51	2.2–2300
Nickel (mg kg^{-1})	7–4300	60–260	19–710	20–310
Lead (mg kg^{-1})	98–14000	5300–26000	2500–10000	3300–22000
Zinc (mg kg^{-1})	610–7800	9000–70000	7000–20000	8100–53000
PCDD (ng g^{-1})	0.25–0.48	115–1040	0.7–32	–
PCDF (ng g^{-1})	0.54–0.102	48–280	1.4–73	–
I-TEQ (ng g^{-1})	0.0018–0.002	1.5–2.5	0.8–2.0	–

Source: IAWG, 1997.[44]

Ash Residues

The ash residues produced from the mass burn incineration of municipal solid waste arise from a number of sources. The ash discharged from the bottom of the grate is known as grate ash and together with the ash material that falls through the grate or grate riddlings is known as bottom ash. Ash also arises in the form of flyash collected from the heat recovery section of the boiler as boiler ash and in the form of air pollution control (APC) ash from systems such as electrostatic precipitators and fabric filters. In addition, the wet scrubber system may also produce a solid phase or sludge material.

Bottom ash from the furnace grate represents the bulk (75–90%) of total ash and is a heterogeneous mixture of slag, ferrous and non-ferrous metals, ceramics, glass, other non-combustible material and uncombusted organic material. The bottom ash consists mainly of silicates, oxides and carbonates. Various compounds and mineral species have been identified in bottom ash, for example $Ca_2Al_2SiO_7$, $Ca_2MgSi_2O_7$, SiO_2, Fe_3O_4, Fe_2O_3, $CaCO_3$, $MgCO_3$, $Ca(OH)_2$, $CaSO_4$, NaCl and KCl and elemental Fe, Al and Cu.[45] The flyash and APC residues are fine particulate material. Table 11 shows the composition for bottom ash, flyash and APC residues from a dry/semi-dry control system and a wet control system.[44] The chemical characteristics of flyash are greatly influenced by

[45] G. Pfranf-Stotz and J. Reichelt, *Characterisation and Valuation of Municipal Solid Waste Incineration Bottom Ashes*, in 1st International Symposium on Incineration and Flue Gas Treatment Technologies, Sheffield, July, 1997.

the additives used in the air pollution control system. For example, where additives such as lime and activated carbon are added, the air pollution control residue would comprise flyash, lime, activated carbon and the associated adsorbed pollutants. In some cases, the lime may comprise up to 50wt% of the APC residue. Flyash is characterised by spherical particles associated with aggregates of polycrystalline, amorphous and glassy material.

Attempts to identify possible mineral phases in the flyash show that, apart from the amorphous and glassy material, various chemical compounds and mineral species can also be identified. The spherical particles common in flyash are composed of complex calcium, sodium and potassium aluminosilicates whilst the associated amorphous and crystalline material is enriched in the more volatile elements.[46] Compounds and minerals identified in flyash include Pb_3SiO_5, $Pb_3O_2SO_4$, $Pb_3Sb_2O_7$, $PbSiO_4$, $Cd_5(AsO_4)_3Cl$, $CdSO_4$, K_2ZnCl_4, $ZnCl_2$, $ZnSO_4$, Fe_3O_4, Fe_2O_3, SiO_2, $CaSiO_3$, Al_2SiO_5, $Ca_3Si_3O_9$, $CaAl_2SiO_6$, $Ca_3Al_6Si_2O_{16}$, $NaAlSi_3O_8$ and $KAlSi_3O_8$.[46–49]

The high heavy metal concentrations present in the ash residues from incineration become of more significance when they are placed in landfill sites where leaching of the pollutants may be a source of groundwater contamination. Generally the flyash is more readily leached than the bottom ash fraction since the heavy metals largely occur in the smallest size fraction of less than 10 μm and are concentrated at or near the surface of the particles.[50] In addition, the relatively high chlorine content of the waste results in the majority of the metal species being present as the metal chlorides, which are generally more soluble in water than other species.[24,51] It has been shown that up to 32.5% of the available zinc, 1.75% of lead, 5.7% of manganese and 94% of the available cadmium can be leached from flyash.[50] Whilst water in contact with flyash produces alkaline solutions rather than acidic, copper, lead, zinc and cadmium show increased solubilities at high alkalinities, that is, they are amphoteric in nature, showing significant solubilities at both low and high pH values.[51] The presence of PCDD and PCDF in the ash samples shown in Table 11 highlights the *de novo* synthesis route to their formation, a route catalysed by the flyash itself. Whilst flyash contains significant concentrations of PCDD and PCDF, bottom ash concentrations are negligible since bottom ashes are quickly quenched before significant PCDD and PCDF production can take place.

The contamination of flyash with heavy metals and PCDD and PCDF result in the ash attaining the status of special or hazardous waste and consequently requiring special permits for landfill disposal.[52,53] Other treatment methods used

[46] T. T. Eighmy, J. D. Eusden, J. E. Krzanowski, D. S. Domingo, D. Stampfli, J. R. Martin and P. M. Erikson, *Environ. Sci. Technol.*, 1995, **29**, 629–646.

[47] C. S. Kirby and J. D. Rimstidt, *Environ. Sci. Technol.*, 1993, **27**, 652–660.

[48] T. Hundesrugge, K. H. Nitsch, W. Rammensee and P. Schoner, *Mull Abfall*, 1989, **6**, 318–324.

[49] J. Evans and P. T. Williams, *Trans. I. Chem. E. Part B*, 2000, **78**, 40–46.

[50] A. Buekens and J. Schoeters, *Thermal Methods in Waste Disposal*, Study performed for the EEC under contract EC1 1011/B721/83/B, A. Buekens, Free University of Brussels, Brussels, 1984.

[51] R. A. Denison and E. K. Silbergeld, *Risk Anal.*, 1988, **8**, 343–355.

[52] E. Mulder, *Waste Manage.*, 1996, **16**, 181–184.

[53] S. Sakai, S. E. Sawell, A. J. Chandler, T. T. Eighmy, D. S. Kosson, J. Vehlow, H. A. van der Sloot, J. Hartlen and O. Hjelmar, *Waste Manage.*, 1996, **16**, 341–350.

or under investigation to stabilise incinerator flyash are solidification, chemical stabilisation, ash melting or vitrification and extraction/recovery processes.[53]

Bottom ash is typically disposed of to landfill or recycled for aggregate use in the construction industry and road building; this is common in Europe.[53] For example, in Germany approximately 50% of bottom ash is utilised, through, for example, sub-base paving applications. In The Netherlands about 90% of bottom ash is used in construction for granular base, or in-fill road base, embankments and noise and wind barriers and it has also been used as aggregate in asphalt and concrete. The main use in Denmark is for development of granular sub-base for car parking, bicycle paths and paved and un-paved roads *etc.*[6]

4 Other Waste Treatment Processes

Whilst waste landfill and incineration of waste remain by far the main options for waste treatment and disposal throughout the world, other established and novel routes are available. Amongst these are such options as, composting, pyrolysis/gasification and recycling of different types of waste components. Such methods are regarded as being more environmentally sustainable and superior to landfill and incineration in relation to the hierarchy of waste management. However, each should be considered in its own right in terms of the emissions from the process.

Composting

Composting of waste involves the aerobic digestion of the biodegradable fraction of municipal solid waste, such as garden and food waste, paper and cardboard *etc.* The biodegradable fraction of municipal solid waste can compose up to 60% of the total mass.[18] A typical aerobic composting system for organic waste includes a pre-processing stage, the aerobic biodegradation stage and a maturation stage. Pre-processing of 'green waste', *i.e.* garden waste from parks and civic amenity sites, may only require shredding or pulverisation. Mixed municipal solid waste would require separation of the component waste on the scale of a materials recovery facility to remove the inert materials such as glass, ferrous and non-ferrous metals *etc.* The composting stage involves the biodegradation of the sample under aerobic conditions and therefore requires aeration of the waste, and is achieved by regularly turning the composting waste or by air injection. Forced aeration systems involve air being blown or sucked through the pile of composting waste by a fan. If the air is drawn down through the pile the odours from the compost are contained in the system allowing for control and treatment if required. Other systems involve the processing of the waste in mechanical systems which allow closer control of temperature, moisture, aeration and waste mixing rates.[54] The maturation stage involves further biodegradation of intermediate compounds and may take several weeks for completion. The final stages of composting would be processes to remove uncomposted materials and contaminants such as glass, plastics and metals and size reduced and screened.

[54] L. F. Diaz, G. M. Savage, L. L. Eggerth and C. G. Golueke, *Composting and Recycling of Municipal Solid Waste*, Lewis Publishers, Boca Raton, USA, 1993.

P. T. Williams

Table 12 Maximum emissions of selected volatile organic chemicals from direct sampling of waste and compost from municipal solid waste composting facilities

Component	Maximum emission (μg m^{-3})	Limit value[1] (μ m^{-3})
Trichlorofluoromethane	915 000	5 620 000
Acetone	166 000	1 800 000
Carbon disulfide	150	31 000
Methylene chloride	260	174 000
2-Butanone	320 000	590 000
Chloroform	54	49 000
1,1,1-Trichloroethane	15 000	1 900 000
Carbon tetrachloride	290	31 000
Benzene	700	32 000
Trichloroethene	1300	270 000
2-Hexanone	6600	20 000
Toluene	66 000	188 000
Tetrachloroethene	5600	339 000
Chlorobenzene	29	46 000
Ethylbenzene	178 000	434 000
m,o-Xylene	15 000	434 000[2]
p-xylene	6900	434 000[2]
Styrene	6100	213 000
Isopropylbenzene	370	246 000
1,3,5-Trimethylbenzene	2200	123 000[2]
1,2,4-Trimethylbenzene	1000	123 000[2]
1,4-Dichlorobenzene	90	451 000
Naphthalene	1400	52 000

[1] American Conference of Governmental Industrial Hygienists, Threshold Limit Values-Time Weighted Average for workplace air.
[2] Value for the sum of all the isomers of each compound
Source: Eitzer, 1995.[56]

The degraded product is a stabilised product which is added to soil to improve soil structure especially for clay soils, acts as a fertiliser improving the nutrient content, acting as a mulch and is used to retain moisture in the soil.

Composting is a biodegradation process and leads to the formation not only of compost but other products which may require control and treatment. The composting of waste involves a range of complex biological processes.[55] Leachate may form in cases of high moisture content. The leachate will have many of the properties and composition of leachate generated in the early stages of landfill. The leachate is allowed to collect in channels and is discharged to sewer or treated on site depending on the level of leachate generated. Gaseous emissions from the composting process consist mainly of volatile organic compounds. The emissions are often malodorous and potentially toxic. Table 12 shows some gaseous emissions from the composting of municipal solid organic waste using a variety of composting systems from windrows, forced aeration, to reactor systems.[56] The results represent maximum observed concentrations and

[55] M. de Bertoldi, G. Vallini and A. Pera, *Waste Manage. Res.*, 1983, **1**, 157–176.
[56] B. D. Eitzer, *Environ. Sci. Technol.*, 1995, **29**, 896–902.

Table 13 Average heavy metal concentrations found in various composted wastes

Heavy metal	EU eco-label standard	Mixed refuse compost	Green waste compost	Source segregated household waste compost
Zinc	300	1510	214	290
Lead	140	513	87	87
Copper	75	274	37	47
Mercury	1.0	2.4	0.4	0.4

Source: Walker, 1996.[58]

in most cases represent localised high concentrations since gas samples were taken directly from the waste and compost piles. However, the results showed that whilst a wide range of emissions were detected they were below permissible workplace exposure limits.

A further aspect of airborne emissions from the composting process is the likely presence of microbiological organisms. The putrescible rich fraction of municipal solid waste has been composted and airborne emissions monitored for bacteria, fungi and actinomycetes.[57] Before composting, moulds and enteric bacteria were found to be the most common microbial species present in the air emissions from mechanically separated waste. During the composting process, the main organisms were fungi and actinomycetes, both of which produce large numbers of spores and can lead to acquired allergic responses. During processing after composting, airborne particles increased markedly, with identified species including *Aspergillus fumigatus* and *Penicillilium* spores, both of which are known to cause allergies.

The final compost product may also contain contaminants and the concentration of heavy metals in compost are subject to regulation. For example, Table 13 shows the concentration of heavy metals found in compost derived from various wastes compared to the European Union standard.[58]

Pyrolysis/Gasification

Pyrolysis/gasification systems for processing waste have a number of advantages over conventional incineration or landfilling of waste. Depending on the technology, the waste can be processed to produce not only energy, but also gas or oil products for use as petrochemical feedstocks and/or a carbonaceous char for use in applications such as effluent treatment or for gasification feedstock.

Pyrolysis processes involve the thermal degradation of organic waste in the *absence of oxygen* to produce three products – a carbonaceous char, oil and combustible gases, all of which may be recycled or used for process energy. The process conditions are altered to produce the desired char, gas or oil end-product, with pyrolysis temperature, heating rate and residence time having the most influence on the product distribution. Gasification of waste is the reaction of oxygen in the form of air, steam or oxygen at high temperature with the available

[57] H.A. Newport, R.P. Bardos, K. Hensler, E. Goss, S. Willett and King, *Municipal Waste Composting*, Report No. CWM/074/93, DoE, HMSO, London, 1993.
[58] M. Walker, *The Waste Manager*, 1996, February, 19–21.

carbon in the waste to produce a gas product, ash and a tar. Partial combustion occurs to produce heat and the reaction proceeds exothermically to produce a low to medium calorific value fuel gas. The operating temperatures are usually relatively high at 800–1100 °C.

A number of systems are available, particularly in the medium to larger scale, for the processing of municipal solid waste, which use combined pyrolysis/gasification. The target product is a synthesis gas which is generally cleaned to some extent and used for power generation. In such processes, an initial waste pyrolysis stage produces a char, oil and gas. In some cases all these products are then transferred to a second gasification stage or the char may be selectively removed for use in other applications. The processes also may have the option of further cleaning of the gas so that it may be used as a chemical feedstock. Such systems are designed to thermally process the waste with the aim of producing no environmentally hazardous emissions.

For example, the Thermoselect system of Switzerland is marketed as a high temperature recycling, closed loop or zero emissions process, where all of the waste is processed to produce a clean synthesis gas, minerals, metals, sulfur, salts, and water.[59] The system uses untreated waste which is compacted at pressures of over 1000 tonnes to 10% of its original volume to form plugs of waste devoid of air. The compacted waste is fed to a pyrolysis reactor heated indirectly at 600 °C. The resultant organic pyrolysis gases, vapours and char are fed to a high temperature gasification chamber operated at approximately 1200 °C with oxygen as the gasifying agent. The product synthesis gas is shock quenched to 90 °C and undergoes several cleaning steps to produce a clean gas suitable as a chemical feedstock or for energy recovery applications. At the base of the gasification reactor, temperatures of 2000 °C are generated through natural gas with oxygen combustion which produces melting of the metal and mineral components of the waste. The liquid melt flows to a homogenisation chamber at 1600 °C where sufficient residence time allows the separation of two phases, a metal alloy and a mineral phase. Rapid quenching of the melt produces a granulate mineral material for use in road building, construction and aggregates, and a metal alloy for recovery of metals. Other products recovered are sulfur from the gas cleaning process and zinc and lead concentrate, sodium chloride and purified water from the water treatment process. The high temperatures involved in the process and rapid cooling of the gases ensure that very low levels of dioxins and furans are reported from the process emissions at levels of $0.002 \, \text{ng m}^{-3}$. It is reported that from one tonne of waste and about 500 kg of oxygen, 890 kg of high purity synthesis gas, 350 kg water, 230 kg vitrified minerals, 29 kg ferrous metals, 10 kg sodium chloride, 3 kg of non-ferrous metal concentrate and 2 kg of sulfur are produced.[59]

Recycling

Recycling is the collection, separation, clean-up and processing of waste materials to produce a marketable material or product. The advantages of using recyclable

[59] B. Calaminus and R. Stahlberg, *Waste Manage.*, 1998, **18**, 547–556.

Table 14 Energy consumption and emissions; comparison of recycled paper with virgin paper production

Source	Recycled paper (per tonne produced)	Virgin paper (per tonne produced)
Energy consumption (GJ)	14.4	22.7
Air emissions (g)		
Particulate	357	4346
CO	383	3165
NO_x	2295	5114
N_2O	280	345
SO_x	6054	10868
HCl	0	4
HF	0.004	0.01
H_2S	0	15
HC	4195	6258
NH_3	2.9	3.4
Hg	0	0.004
Water emissions (g)		
BOD	1	2921
COD	3	25423
Suspended solids	1	1
TOC	25	30
AOX	0	3
Ammonium	0.331	0.876
Chloride	9	22
Fluoride	0.714	1.89
Sulfide	0	7
Solid waste (kg)		
Total solid waste	70.6	150.2

Source: White *et al.*, 1995.[15]

materials are that there is reduced use of virgin materials with consequent environmental benefits in terms of energy savings in the production process, reduced emissions to air and water and onto land. However, recycling may not always be the best environmental option for a particular type of waste and a full analysis of the processes involved in recycling *versus* treatment and disposal should be made. Such analyses may be undertaken using life cycle analysis. Life cycle analysis is the analysis of a product throughout its lifetime to assess its impact on the environment. The analysis quantifies how much energy and raw materials are used and how much solid, liquid and gaseous waste is generated at each process stage of the product's life. Life cycle analyses comparing recycling *versus* manufacturing of the product from virgin materials have been used to highlight the benefits of recycling. For example, Table 14 shows a comparison of a life cycle assessment for recycled paper *versus* virgin paper using data from Swedish and Swiss sources.[15] In most cases, the emissions from the recycling process for paper are less than from the production of virgin paper. However, this

P. T. Williams

may not always be the case for the production of recycled products; for example, it has been suggested that the recycling of glass produces higher solid waste emissions than the production of virgin glass.[15] But it should also be noted that emissions to air and water for glass recycling are lower than for virgin glass production.

Health Impacts of Waste Incineration*

ARI RABL AND JO V. SPADARO

1   Introduction

Incineration is a major option for the disposal of waste, advantageous because it greatly reduces the space requirement for landfills yet highly controversial because of perceived health risks from air pollution, especially dioxins. Whereas such fears are natural after the bad practices of the past, the decisions to be considered now concern new technologies, subject to stringent regulations such as the new Directive 2000/76/EC.[1] With new technologies the emission of pollutants is orders of magnitude lower than with the old unregulated incinerators that had little or no flue gas clean up. In passing we also note that other thermal treatment technologies (pyrolysis and gasification) are being developed that can have even lower impacts.

In the present paper we offer a perspective on the health impacts of air pollution from incinerators, by making a number of comparisons, in terms of concentrations, damages and damage costs. We do not attempt a full analysis of

---

* Most of this paper is drawn from Rabl, Spadaro and McGavran,[2] with permission from *Waste Management and Research*, to which the reader is referred for more complete documentation; the numbers are essentially unchanged, except for dioxin impacts which are based on the more recent EPA.[3] Some of the material is from Rabl and Spadaro,[4] with permission from the *Annual Review of Energy and the Environment*, Vol. 25, © 2000 by Annual Reviews.

[1]   EC, Directive 2000/76/EC of the European Parliament and of the Council of 4 December 2000 on the incineration of waste.
[2]   A. Rabl, J. V. Spadaro and P. D. McGavran, Health risks of air pollution from incinerators: a perspective, *Waste Manage. Res.*, 1998, **16**, 365–388. This paper received the 1998 ISWA (International Solid Waste Association) Publication Award.
[3]   EPA, *Exposure and Human Health Reassessment of 2,3,7,8-Tetrachlorodibenzo-p-Dioxin (TCDD) and Related Compounds: Part III: Integrated Summary and Risk Characterization for 2,3,7,8-Tetrachlorodibenzo-p-Dioxin (TCDD) and Related Compounds*, Report EPA/600/P-00/001Bg, United States Environmental Protection Agency. Washington, DC 20460, September 2000, available at www.epa.gov/ncea.
[4]   A. Rabl and J. V. Spadaro, Public health impact of air pollution and implications for the energy system, *Annu. Rev. Energy Environ.*, 2000, **25**, 601–627.

Issues in Environmental Science and Technology, No. 18
Environmental and Health Impact of Solid Waste Management Activities
© The Royal Society of Chemistry, 2002

the problem of waste treatment which should involve a comparison with alternative treatment options, but we provide elements for such an analysis. Also we do not address liquid or solid residues, other than to emphasize the need for good management.

Compared to land filling, incineration can greatly reduce the quantity of organic pollutants. As for heavy metals, while their quantity does not change, their form and toxicity do; however, since the residues from incineration are much smaller in volume and more homogeneous than the original waste, it is much easier to ensure proper disposal in well-managed special landfills or stabilization in concrete. The reduction of waste volume is about a factor of 4–10, if one does not take into account the nature of the residues remaining after incineration. But most of the residue consists of bottom ash or slag, relatively harmless because the toxic metals tend to concentrate in the fly ash instead. Incinerator slag can be used as road fill, requiring only minor treatment. Fly ash can be stabilized in concrete, with negligible risk of lixiviation.[5]

If most municipal solid waste (MSW) is put into landfill, the land requirement is so large that sooner or later the land will be reused for other purposes, with attendant health risks for future generations. With incineration, by contrast, the health risks are borne mostly by the generation that produced the waste.

We quantify, as far as data and models are available, the damage due to air pollution by carrying out an analysis of impact pathways (environmental fate) whose principal steps are the following:

- specification of the emissions (*e.g.* kg s^{-1} of particles emitted by stack);
- calculation of increased pollutant concentrations in all affected regions (*e.g.* μg m^{-3} of particles, using models of atmospheric dispersion);
- calculation of damages (*e.g.* number of asthma attacks due to these particles, using a dose–response function*);
- monetary valuation of this damage (*e.g.* multiplication by the cost of an asthma attack).

The resulting values are summed over all receptors (population, plants, buildings,…) that are affected, choosing the temporal and spatial boundaries of the analysis such as to ensure that essentially all the damage is taken into account. We draw on the major studies of damage costs of pollution in the USA[6,7] and Europe;[8] the latter, known as the ExternE Project of the European Commission,

---

*For air pollutants also known as exposure–response or concentration–response (CR) functions.

[5] B. Germaneau, B. Bollotte and C. Defossé, *Leaching of Heavy Metals by Mortar Bars in Contact with Drinking and Deionized Water, Emerging Technologies Symposium on Cement and Concrete in the Global Environment*, Portland Cement Association, 10–11 March 1993, Chicago, IL.

[6] ORNL/RFF, *External Costs and Benefits of Fuel Cycles*, Prepared by Oak Ridge National Laboratory and Resources for the Future, ed. R. Lee, Oak Ridge National Laboratory, Oak Ridge, TN 37831, 1994.

[7] R. D. Rowe, C.M. Lang, L. G. Chestnut, D. Latimer, D. Rae, S. M. Bernow and D.White, *The New York Electricity Externality Study*, Oceana Publications, Dobbs Ferry, New York, 1995.

[8] ExternE, *ExternE: Externalities of Energy*. Vol. 7: *Methodology 1998 Update* (EUR 19083); Vol. 8: *Global Warming* (EUR 18836); Vol. 9: *Fuel Cycles for Emerging and End-Use Technologies, Transport and Waste* (EUR 18887); Vol. 10: *National Implementation* (EUR 18528). Published by European Commission, Directorate-General XII, Science Research and Development. Office for

is continuing and the authors are active participants. Since all of these studies have found the damage costs of air pollution to be dominated by health impacts, we focus on health in the present paper.

To begin we comment on several key issues. The first is site dependence: a ton of SO_2 emitted in a big city causes severe health damage, but emitted in the middle of the ocean its health damage is negligible. Site dependence is explicitly taken into account in the impact pathway analysis. This realism is an awkward complication for the presentation of the results and their use for public policy. How could one obtain results that are 'typical' and draw general conclusions?

To set the stage for general conclusions, we show results calculated by detailed site-specific modelling and compare them with a simple formula (the 'uniform world' model) that follows from conservation laws in the limit of uniform receptor density. We find the simple formula instructive because it gives an order of magnitude estimate; for primary pollutants the true damage can be about three times smaller or six times larger depending an stack height and type of site (rural *vs.* urban). For secondary pollutants the variability with site is much weaker, about $\pm 50\%$ for nitrates and sulfates (for the same average population density).

The second key issue concerns epidemiology: the crucial assumption for our calculations of health impacts is linearity of incremental damage with incremental exposure. In reality the dose–response functions may well have thresholds and curvature, but there are insufficient data to justify non-linear models for our calculations. Linearity is commonly assumed for carcinogens (and justified by the data for some although not all carcinogens).[9] Linearity is also indicated for population level mortality impacts of the classical air pollutants, especially PM.[10,11] Note that for the calculation of incremental damage a threshold has no effect if it is below background exposure: a 'hockey stick' gives the same results as a line through the origin with the same slope.

Linearity has important implications for public policy. It directs the analysis toward the total population dose rather than peak individual doses, at least as far as new clean technologies are concerned because their emissions are so low that even the highest individual dose does not entail any significant augmentation of individual risk; also the analysis must extend over large regions to capture most of the damage. Furthermore, linearity implies that there is no safe level for incremental emissions. Since the cost of pollution control is high and increases strongly as emissions are reduced further, rationality calls for risk comparisons and cost–benefit analysis.

The third key issue concerns monetary valuation. Monetization greatly simplifies the presentation of the results because it converts a large number of

Official Publications of the European Communities, L-2920 Luxembourg, 1998. Results are also available at http://ExternE.jrc.es/publica.html.

[9] C.H. Frith, N.A. Littlefield and R. Umholtz, Incidence of pulmonary metastases for various neoplasms in BALB/cStCrlfC3H/Nctr female mice fed N-2-fluorenylacetamide, *J. Nat. Cancer Inst.*, 1981, **66**, 703–712.

[10] M.J. Daniels, F. Dominici, J.M. Samet and S.L. Zeger, Estimating particulate matter–mortality dose–response curves and threshold levels: an analysis of daily time-series for the 20 largest US cities, *Am. J. Epidemiol.*, 2000, **152**, 397–406; see also Comment in: *Am. J. Epidemiol.*, **152**, 407–412.

[11] C.A. Pope, Invited commentary: particulate matter–mortality exposure response relations and threshold, *Am. J. Epidemiol.*, 2000, **152**, 407–412.

incommensurate impacts to a common unit. However, in view of the controversies surrounding the monetary valuation of mortality, some people prefer to refrain from an economic valuation and use instead a multicriteria analysis. For that purpose we point out that about 85% of our health damage cost estimates is due to premature mortality; thus the damage costs are approximately proportional to the years of life lost.

Going down the steps of an impact pathway analysis, the results come progressively closer to criteria of direct concern, but involve progressively more assumptions and uncertainties. It is therefore advisable to consider a gamut of different comparisons. As illustration we consider a scenario where all municipal solid waste (MSW) is incinerated with emissions equal to the limit values of the EC Directive.[1] We offer the following comparisons, to the extent that suitable data are available:

- incremental concentration (or dose) compared to background concentration (or dose);
- incremental concentration (or dose) compared to health guidelines (EC or WHO);
- health risks from different pollutants compared to each other;
- incremental damage cost compared to the cost of incineration itself.

Rabl, Spadaro and McGavran[2,*] provide two additional comparisons, based on the same assumptions,

- incremental emission compared to other emissions (natural and anthropogenic, *e.g.* cars);
- incremental years of life lost compared to other risks of everyday life.

The comparisons of emissions and of concentrations are instructive because they show whether MSW is significant relative to other pollution sources that one may or may not be able or willing to reduce. It also has the great advantage of being unaffected by the dominant source of uncertainty, namely epidemiology.

2 Emissions

Pollutants

In terms of total environmental costs the greenhouse emissions, especially carbon dioxide (CO_2) and methane (CH_4), are likely to dominate. However, most of their impacts are in categories other than direct health effects. Furthermore, an inventory of relative greenhouse gas emissions for alternative MSW treatment options would require fairly detailed specification of the landfill or other alternatives, in particular how much of the CH_4 from the decay of organic matter is captured. That is beyond the scope of the present paper, and so we limit our discussion to the other air pollutants.

We consider the 'classical air pollutants': the primary pollutants particulate matter (PM), nitrogen oxides (NO_x), sulfur oxides (SO_2), volatile organic

---

*The damage costs in the present paper are different from those in our earlier publications because the methodology has been evolving, but, apart from dioxin, the differences are negligible.

compounds (VOC) and carbon monoxide (CO), as well as the secondary air pollutants nitrates, sulfates and ozone (O_3). In addition we discuss heavy metals and dioxins (the 'micropollutants'). We pay special attention to the latter because fear of dioxins is one of the main reasons for vehement opposition to most waste incineration projects.

Dioxin is a name for a family of 75 chlorinated aromatic compounds, to which one might add 135 closely related compounds, the polychlorinated dibenzofurans. Several of these are highly toxic, the most toxic and carcinogenic being 2,3,7,8-tetrachlorodibenzo-*p*-dioxin, usually abbreviated TCDD. Dioxins and furans can be produced in trace quantities during the combustion of chlorinated organic compounds. They are destroyed by exposure to light within hours, but in the soil they may persist for more than ten years.[12] Implicated in the Seveso accident 1976 and blamed for health effects from the defoliation with Agent Orange in Vietnam, dioxins have acquired a frightful reputation.

Dioxins are not produced intentionally; rather, they are an undesirable byproduct of certain industrial processes. An important source is incineration of chlorinated plastics – which is one of the reasons why there has been so much opposition to waste incineration. But sources of dioxins are ubiquitous, and even the burning of ordinary wood produces some. Thus it comes as no surprise to find evidence for dioxins in preindustrial times (ref. 13, vol. II, p. 3.146). It is convenient to state all dioxin data as TEQ = toxic equivalent 2,3,7,8-TCDD, and we will do so throughout this paper.

Data and Regulations

Emissions depend on the composition of the waste and on the equipment used to treat the waste, especially the flue gas clean up; for a given installation the emissions can vary with varying operating conditions. Figure 1 shows the concentration of pollutants in the flue gas, as measured for older plants, and compares them with regulations in the US and Europe. The concentrations are expressed in terms of m^3 at 273 K and 101.3 kPa. The data for older plants are mostly based on data from the UK as reported by Williams.[14] The regulations in the USA have been in effect for new incinerators since 1991. In the European Union the EC Directives of 1989 have provided comparable regulation,[15] now replaced by the new Directive 2000/76/EC,[1] to take effect by the end of 2002. The limit values of the latter are indicated in the labels of Figure 1.

The dioxin data in Figure 1 are expressed as TEQ (toxic equivalent of TCDD), assuming TEQ = total dioxin/60 as typical value for waste incinerators.[13] For MSW incinerators in the USA EPA indicates an average dioxin emission of 0.0000001 g kg_{waste}^{-1}, converted to 20×10^{-6} mg m^{-3} by assuming 5.15 m^3

[12] F. H. Tschirley, Dioxin. *Sci. Am.*, 1986, **254** (Feb.), 29.

[13] EPA, *Estimating Exposure to Dioxin-like Compounds*, Report EPA/600/6-88/005Ca, b and c, United States Environmental Protection Agency, Washington, DC 20460, June 1994.

[14] P. T. Williams, *Pollutants from Incineration: an Overview*, in *Waste Incineration and the Environment*, ed. R. E. Hester and R. M. Harrison, Issues in Environmental Science and Technology No. 2, Royal Society of Chemistry, Cambridge, 1994, pp. 27–52.

[15] EC, Directives 89/369/EEC and 89/429/EEC on the incineration of municipal waste.

A. Rabl and J. V. Spadaro

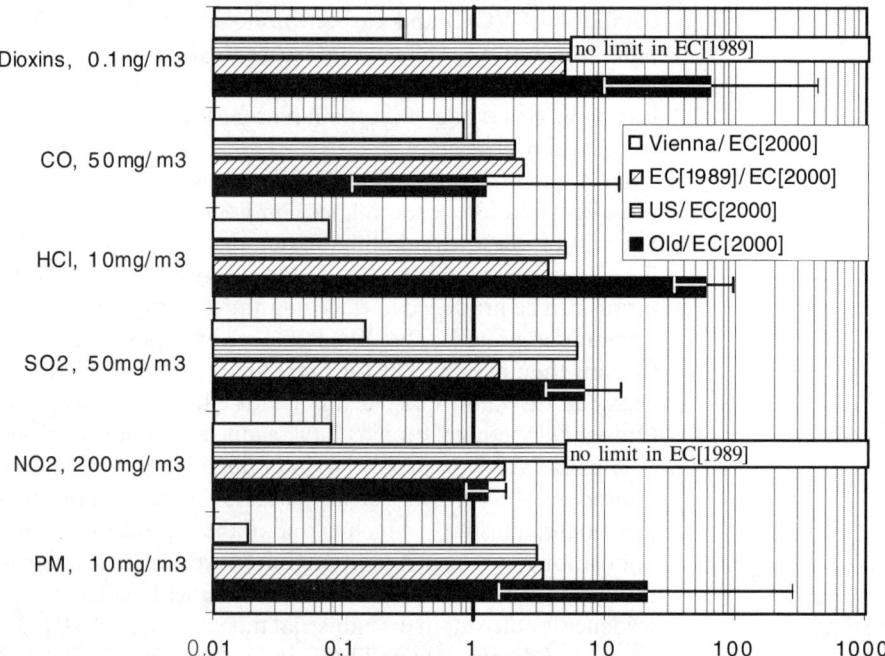

Figure 1 Emissions of MSW incinerators. The limit values of the EC Directive[1] are indicated in the labels. The bars show the emissions in units of the EC limit values, on logarithmic scale. Measured data for older incinerators are mostly for UK data, as cited in Table 1 of Williams,[14] the error bars indicating range of values. US = New Source Performance Standards in USA, as cited in Steverson.[18] Vienna = measured data for MSW incinerator of Vienna, Austria.

kg_{waste}.[16] For industrial waste incinerators built before 1980 Table 4 of Jones *et al.*[17] shows $1-7 \times 10^{-6}$ mg m^{-3} TEQ.

We do not show toxic metals in Figure 1 because different regulations are formulated in terms of different combinations of metals. The limit values of EC[1] are listed in Table 1, together with a typical breakdown according to ETSU.[16] In the past much of the Hg and Pb came from batteries, a contribution now greatly reduced thanks to recycling (especially of Pb in car batteries) and elimination at the source (Hg in household batteries).

The annual average of real emissions is lower than the limit values (assuming that the regulations are enforced) because the operator of an incinerator must allow for a sufficient safety margin to cope with fluctuations. The margin depends on the pollutant and on the installation. Lacking data for new incinerators, we assume emissions equal to the limit values of EC.[1]

There are technologies that are already much better than the limit values of EC.[1] A particularly interesting one is the MSW incinerator in Vienna which has been operating in its present version since 1992.[19] It is a showcase for clean technology, from the sophisticated clean up equipment (emissions monitored and

[16] ETSU, *Economic Evaluation of the Draft Incineration Directive*, Report for the European Commission DG11, ETSU, Harwell Laboratory, Didcot, Oxfordshire OX11 0RA, UK, 1996.

[17] P.H. Jones *et al.*, The global exposure of man to dioxins: a perspective on industrial waste incineration, *Chemosphere*, 1993, **26**, 1491–1497.

[18] E.M. Steverson, *The US Approach to Incinerator Regulation*, in *Waste Incineration and the Environment*, ed. R.E. Hester and R.M. Harrison, Issues in Environmental Science and Technology No. 2, Royal Society of Chemistry, Cambridge, 1994, pp. 113–135.

[19] Wien, Spittelau District Heating Plant: a Description of the Works and Functions, Fernwärme Wien, Spittelauer Lände 45, A-1090 Wien 9, 1995.

Table 1 Limit values of EC[1] for toxic metals. Where the limit values are given as sum of emissions for a group of metals, a typical breakdown is indicated according to data of ETSU.[16]

Metal or group of metals	Limit values, mg m^{-3}
Hg	0.05
Cd + Tl	0.05
Cd = 81% of sum of Cd and Tl	
As + Co + Cr + Cu + Mn + Ni + Pb + Sn + Sb + V	0.5
As = 3%, Cr = 6%, Ni = 34% and Pb = 22% of sum of As, Co, Cr, Cu, Mn, Ni, Pb, Sn, Sb and V	

publicly displayed in real time) to the exterior (designed by the painter and architect Hundertwasser). Measured data for this installation are included in Figure 1.

Even better performance can be achieved with various versions of pyrolysis. An example is the one developed by Siemens and called thermolysis, with several features that minimize emission of pollutants and production of residues.[20] The fly ash trapped by the electrofilter is sent back into the combustion chamber where it becomes part of the slag. The gas leaving the pyrolysis unit is burned at 1300°C to minimize the formation of dioxins. Liquid emissions are zero because wastewater is evaporated. The process has reached technical maturity with a unit in Fürth, Germany, which has begun operations with a capacity of 100 000 000 kg yr^{-1}. Clearly there has been impressive progress in pollution control.

Another interesting technology is the co-incineration of wastes in cement kilns, as replacement of conventional fuels. Even though only certain wastes of sufficiently constant composition are suitable, for instance used oils, used solvents, and shredded tires, cement kilns can make a very significant contribution to the management of the waste problem. They are especially well suited for toxic organic wastes because the temperatures are higher and residence times longer (5 s between 1500 and 2000°C) than in most specialized waste incinerators (2 s at 1200°C). Furthermore, the fly ash is added to the raw materials and will thus be immobilized in the resulting cement. Note that stabilization in concrete is considered one of the best available techniques for the management of fly ash. Measurements have shown that leaching of toxic metals is negligible (well below regulatory limits) even when concrete is doped with toxic metals at concentrations orders of magnitude higher than regular concrete; in particular for Cd and Pb no significant increase in leaching rate was detectable between regular and doped concrete.[5]

Energy Recovery and Avoided Emissions

There is another consideration: avoided emissions by virtue of energy recovery which can render the net emissions attributable to waste incineration much lower than the above values. Of course, net emissions depend on the energy source that is avoided. Obviously, the dirtier the displaced source, the lower the net

[20] B. Mortgat, PTR de Siemens: avènement de l'ère industrielle de la thermolyse des déchets (PTR of Siemens: the arrival of the industrialization of pyrolysis of wastes), *Environnement and Technique*, No. 160, Oct.1996, 61–66.

Table 2 Net emission of pollutants from MSW incineration, assuming emission equal to limit values of EC,[1] with flue gas volume 5.15 N m^3 kg$_{waste}$$^{-1}$, and 2.00 kW h kg$_{waste}$$^{-1}$ heat recovery. Data for avoided emissions from Tabet.[21]

	Incinerator emission g kg$_{waste}$$^{-1}$	If incinerator displaces gas fired boiler or furnace			If incinerator displaces oil fired boiler or furnace @1% S		
		Avoided emission g (kW h)$^{-1}$	Credit per kg$_{waste}$ g kg$_{waste}$$^{-1}$	Net emission g kg$_{waste}$$^{-1}$	Avoided emission g (kW h)$^{-1}$	Credit per kg$_{waste}$ g kg$_{waste}$$^{-1}$	Net emission g kg$_{waste}$$^{-1}$
PM$_{10}$	0.052	0.00	0	0.052	0.11	0.211	−0.16
NO$_2$	1.03	0.24	0.485	0.545	0.70	1.393	−0.363
SO$_2$	0.258	0.00	0	0.258	2.12	4.235	−3.978
CO$_2$	862	219	438	423	324	647	215

emissions. Even though MSW is a dirty fuel, its incineration with new technologies emits less than cleaner fuels with old technologies.

This is illustrated by the examples in Table 2, showing net emissions for three fairly typical conditions:

(a) no energy recovery (2nd column);
(b) the heat from the incinerator displaces a gas fired boiler (5th column);
(c) the heat from the incinerator displaces an oil fired boiler (8th column).

For heat recovery we assume 2.0 kW h kg$_{waste}$$^{-1}$ as typical thermal energy production by incinerators coupled to a district heating system, based on data for MSW incinerators in Paris and in Vienna. Data for avoided emissions are adapted from Tabet.[21] The SO$_2$ emissions are proportional to the sulfur content of the fuel since boilers and furnaces for space heating do not have flue gas desulfurization. Here we assume oil with a sulfur content of 1%. To explain the calculations for Table 2, take particulate matter for which the emission is 0.052 g kg$_{waste}$$^{-1}$. If heat recovery displaces an oil fired boiler that emits 0.11 g (kW h)$^{-1}$, the credit per kg of waste is 0.11 g kW h^{-1} × 2.0 kW h kg$_{waste}$$^{-1}$ = 0.211 g kg$_{waste}$$^{-1}$; the net emission is thus 0.052−0.211 = −0.16 g kg$_{waste}$$^{-1}$ (numbers have been rounded from original spreadsheet).

Emissions Scenario

Since incineration is of greatest interest and concern if it is used on large scale, we consider in this paper the consequences of a scenario where all MSW is incinerated. For the quantities of MSW, the data of Denison and Ruston[22] indicate 250–750 kg yr^{-1} per capita in industrialized countries during the mid-1980s. There is a tendency for waste to increase with general wealth, but this is compensated by source reduction and efforts to recycle. On balance we assume:

[21] J. P. Tabet, Une technologie respectueuse de la qualité de l'air: la cogénération (Cogeneration: a technology that respects air quality), *Les Entretiens de la Technologie*, mars 1996.
[22] R. A. Denison and J. Ruston, *Recycling and Incineration: Evaluating the Choices*, Island Press, Washington, DC, 1990.

$$\text{MSW production} = 500 \text{ kg}_{waste} \text{ yr}^{-1} \text{ per capita} \qquad (1)$$

as a typical round number. To calculate the emissions per kg of waste or per capita, we multiply the limit values of Figure 1 by a typical value of flue gas volume:[15]

$$\text{flue gas volume} = 5.15 \text{ m}^3 \text{ kg}_{waste}^{-1} \qquad (2)$$

3 Dispersion and Peak Concentration

Dispersion Models

For most air pollutants atmospheric dispersion is significant over hundreds to thousands of km.[23] Both local and regional effects are important. We have therefore used a combination of local and regional dispersion models.

To model dispersion over the short range, up to tens of km from the source, the gaussian plume is considered adequate, and we have used the ISC model,[24] a gaussian plume model approved by the EPA. Over such short distances the depletion rates of the pollutants under consideration are negligible; hence the same calculation gives the peak concentration for all pollutants.

At the regional scale we have used two different models, the Harwell Trajectory model as adapted by Krewitt *et al.*[25] for the EcoSense software used by ExternE, as well as the EMEP model of the Norwegian Meteorological Service,[26,27] the model chosen for the official allocation of acid rain budgets among the countries of Europe.

These dispersion calculations have been coupled with an integration over population data, using two software packages developed independently for this purpose: EcoSense and PathWays2.0.[28] Whereas EcoSense includes the Harwell Trajectory Model, for the PathWays2.0 calculations we have used EMEP results. Both sets of calculations use the ISC model for the local dispersion. We have compared the results for total damage per kg of pollutant between these two sets of calculations and found agreement within approximately 20%.[29] Ozone damage due to the precursor NO_x has been calculated by Rabl and Eyre,[30] using the EMEP results of Simpson.[27]

[23] J. H. Seinfeld and S. N. Pandis, *Atmospheric Chemistry and Physics: from Air Pollution to Climate Change*, John Wiley and Sons. New York, 1998.

[24] R. W. Brode and J. Wang, *User's Guide for the Industrial Source Complex (ISC2) Dispersion Model*, Vols. 1–3, EPA 450/4-92-008a, EPA 450/4-92-008b and EPA 450/4-92-008c, US Environmental Protection Agency, Research Triangle Park, NC 27711, 1992.

[25] W. Krewitt, A. Trukenmueller, P. Mayerhofer and R. Friedrich, *ECOSENSE – an Integrated Tool for Environmental Impact Analysis*, in *Space and Time in Environmental Information Systems*, ed. H. Kremers and W. Pillmann, Umwelt-Informatik aktuell, Band 7, Metropolis-Verlag, Marburg, 1995.

[26] H. Sandnes, *Calculated Budgets for Airborne Acidifying Components in Europe*, EMEP/MSC-W Report 1/93, Norwegian Meteorological Institute, PO Box 43, Blindern, N-0313 Oslo 3, July 1993.

[27] D. Simpson, Photochemical model calculations over Europe for two extended summer periods: 1985 and 1989. Model results and comparison with observations, *Atmos. Environ.*, 1993, **27A**, 921–943.

[28] P. S. Curtiss and A. Rabl, *Impact Analysis for Air and Water Pollution: Methodology and Software Implementation*, in *Environmental Modeling – Vol. 3*, ed. P. Zannetti, 1996, Chapter 13, pp. 393–426.

[29] A. Rabl and J. V. Spadaro, Environmental damages and costs: an analysis of uncertainties, *Environ. Int.*, 1999, **25**, 29–46.

[30] A. Rabl and N. Eyre. An estimate of regional and global O_3 damage from precursor NO_x and VOC emissions, *Environ. Int.*, 1998, **24**, 835–850.

Figure 2 Maximum incremental concentration c_{max} due to incinerator, in ng m^{-3}, for an emission rate of 1000 kg yr^{-1}, as function of stack height, calculated with ISC[24] for urban conditions

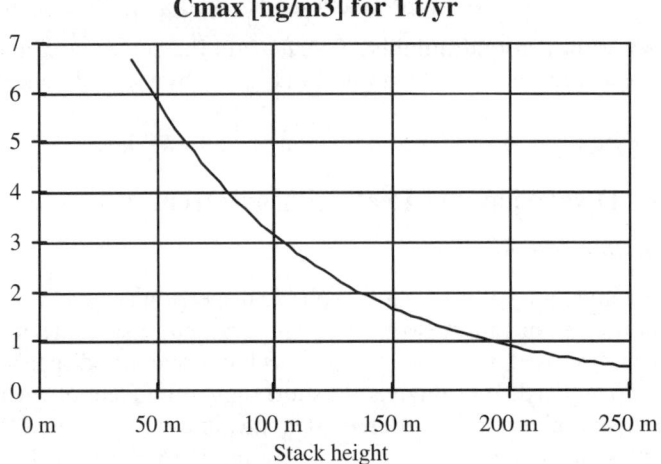

Cmax [ng/m3] for 1 t/yr

Peak Concentrations

Peak concentrations occur within a few km of the source. Figure 2 shows the variation of the peak concentration c_{max} (annual average), calculated by ISC for urban conditions with meteorological data for Paris. The value of c_{max} decreases strongly as the stack height is increased. By contrast to the value of c_{max}, its location does not vary much with stack height; this comes about because of a complicated interplay between stack height, atmospheric stability classes and height of boundary layer. For this calculation Paris appears to be quite a representative choice: the wind speeds are average, and the wind directions are relatively uniform.

An interesting comparison emerges when one combines Figure 2 with the emissions data in Figure 1 and Table 1. Let us consider an MSW incinerator with annual throughput of 0.25E9 kg yr^{-1}, corresponding to a city of half a million with the assumed scenario. If this incinerator has a stack height of 100 m and the emissions equal the limit values of EC,[1] one readily finds the maximum incremental concentration values c_{max} in Table 3.

For comparison with c_{max} we show typical concentrations that have been measured in urban and rural environments. The concentrations for heavy metals are for Europe; note that Pb in Europe for the period of the data is high because unleaded gasoline has become compulsory for new cars only since the beginning of the 1990s.

Column seven, Urban/c_{max}, shows that the highest increment due to the assumed incinerator is small compared to existing backgrounds in urban or industrial environments; this is the case even for Hg which has the lowest ratio Urban/$c_{max} = 4$. For cases where a range is given under 'Urban' we have taken the geometric mean because ambient pollutant concentrations tend to be lognormal.[31]

Finally we compare c_{max} with guidelines for public health; the ratio in the last

[31] W. R. Ott, *Environmental Statistics and Data Analysis*, Lewis Publishers, CRC Press, Boca Raton, FL 33431, 1995.

Table 3 Annual output of pollutants and resulting maximum incremental concentrations c_{max} due to a MSW incinerator with emissions equal to the limit values of EC,[1] with throughput of 0.25E9 kg yr^{-1} and stack height 100 m. For comparison columns 4–6 show typical ambient concentrations, and column 7 shows ratio of c_{max} and geometric mean of urban values. Last two columns show WHO guidelines[32] and their ratio with c_{max}. Blanks = we have not found any data

Pollutant[a]	Due to MSW incinerator		Typical ambient concentrations				Health guidelines		
	kg/yr	c_{max} ng m^{-3}	Remote ng m^{-3}	Urban ng m^{-3}	Industrial ng m^{-3}	Urban/c_{max}	EPA[c] ng m^{-3}	WHO[d] ng m^{-3}	Safety factor WHO/c_{max}
As[b]	20	0.05	0–2	5–50	8–200	295			
Cd[b]	50	0.16	0.1–1	1–50	1–100	45			
Cr[b]	40	0.13	0–3	4–70	5–200	134			
Hg[b]	60	0.19	0.001–6	0.1–5	0.5–20	4		300	1580
Ni[b]	220	0.65	0.1–0.7	3–100	8–20	27			
Pb[b]	140	0.42	0.2–60	80–4000	50–450	1331	1.5×10^3	1×10^3	2350
Cd+Tl	60	0.19							
As+...+V	640	1.93							
Dioxins	1.29×10^{-4}	3.86×10^{-7}	1.0×10^{-4}	>126–5625		>436		4.4×10^{-4}	1300
PM	12 900	39	21×10^3	34×10^3	58×10^3	260	5×10^4	5×10^4	1280
NO$_2$	257 500	773	7×10^3	46×10^3	86×10^3	871	1.0×10^5	1.5×10^5	190
SO$_2$	64 400	193	4×10^3	25×10^3	52×10^3	59	8×10^4	5×10^4	260
CO	64 400	193		1–5×10^6		11.5×10^3	1.0×10^7	1.0×10^7	51 800
HCl	12 900	39							
HF	1300	4							

[a] As+..+V = sum of As, Co, Cr, Cu, Mn, Ni, Pb, Sn, Sb and V; Dioxins as TEQ.
[b] With typical breakdown of metal emissions as estimated by ETSU;[18] see Table 1.
[c] Guidelines by EPA[33] for maximum permissible concentration (annual average, except quarterly for Pb and 8 h for CO).
[d] Guidelines by WHO[32] for maximum permissible concentration; WHO limit for dioxins = 10 pg kg$_{body}$ $^{-1}$ day^{-1} TEQ, converted to
10 pg kg$_{body}$ $^{-1}$ day^{-1} × (2.2/119) × 0.42 m^3/(kg$_{body}$ $^{-1}$day^{-1}) = 4.4×10^{-4} ng m^{-3}.
Sources for concentration data:
for PM, NO$_2$, SO$_2$ data for USA 1985 as cited in Chapter 2 of OECD;[34]
for CO data for France 1993 (Stroebel et al.[35]);
for heavy metals: *Heavy metals: identification of air quality and environmental problems in the European Community*; report EUR 10678 EN/I, Brussels-Luxemburg, as cited in T.5 of OECD;[36] includes only As+Co+Cr+Cu+Mn+Ni+Pb+Sb+V (no Sn);
typical dioxin TEQ concentration is 9.95×10^{-5} ng m^{-3} in North America and 1.08×10^{-4} ng m^{-3} in Europe [EPA,[13] vol. 2, Table 4.11].

A. Rabl and J. V. Spadaro

column of Table 3 can be considered a 'safety factor'. Note that the comparison of c_{max} with EPA and WHO guidelines is independent of the calculation of damages, in the following Sections. If EPA and WHO guidelines correspond to a threshold below which there is no impact, then none of the pollutants covered by the guidelines causes any damage, except at sites where the existing background concentrations are already near or above the guideline: the last column in Table 3 (and analogous ratios for EPA) indicates that the safety factors are quite large, at least two orders of magnitude.

The calculation of damages and costs, by contrast, assumes linearity of incremental damage at current ambient concentrations, in clear contradiction with the interpretation of the guidelines as 'zero risk' values. In view of the uncertainties surrounding CR functions, we propose Table 3 and the calculations in the following sections as complementary perspectives.

4 Health Impacts and Costs

General Remarks

A consensus has been emerging among public health experts that air pollution, even at current ambient levels, is associated with a variety of health problems, especially respiratory diseases and mortality [see *e.g.* refs. 37–39]. However, with epidemiology it is difficult to identify specific causes because people are exposed to a mix of pollutants and the different air pollutants tend to be correlated with each other. Most recent studies have suggested fine particles as a prime culprit; ozone has also been implicated directly. There may also be significant direct health impacts of SO_2, but for direct impacts of NO_x the evidence is less convincing. The uncertainties are large and the risk of double counting when summing the damage costs over pollutants cannot be ruled out.

Depending on the epidemiological approach used to determine a CR function, one talks about acute and chronic CR functions. The most common approach, and the easiest to implement, is to carry out a time series study of a population by identifying short term correlations (over a few days) between air pollution and a health end-point. One chooses a functional form (typically linear, logarithmic or

[32] WHO, *Air Quality Guidelines for Europe*, European Series No. 23. World Health Organisation, Regional Publications, Copenhagen, 1987.

[33] EPA, Subchapter C – Air Programs, Part 50. National Primary and Secondary Ambient Air Quality Standards. Code of Federal Regulations 50: 693–697, United States Environmental Protection Agency, Washington, DC 20460, 1991.

[34] OECD, *OECD Environmental Data Compendium*, 1995.

[35] R. Stroebel, V. Berthelot and B. Charré, *La qualité de l'air en France 1993–94 (Air quality in France 1993–94)*, ADEME, 27 rue Louis Vicat, F-75015 Paris, and Ministère de l'Environnement, 20 ave. de Ségur, F-75007 Paris, 1995.

[36] OECD, *Control of Hazardous Air Pollutants in OECD Countries*, 1995.

[37] F. W. Lipfert, *Air Pollution and Community Health: a Critical Review and Data Sourcebook*, Van Nostrand Reinhold, New York, 1994.

[38] R. Wilson and J. D. Spengler (ed.), *Particles in Our Air: Concentrations and Health Effects*, Harvard University Press, Cambridge, MA, 1996.

[39] R. Bascom, P. A. Bromberg, D. L. Costa, R. Devlin, D. W. Dockery, M. W. Frampton, W. Lambert, J. M. Samet, F. E. Speizer and M. Utell, Health effects of outdoor air pollution, *Am. J. Resp. Crit. Care Med.*, 1996, **153**, 3–50 (Part 1); 477–498 (Part 2).

exponential) with one adjustable parameter (more cannot be identified in practice), and determines the parameter by regression against the pollution data. Time series studies identify only short term effects and yield acute CR functions. This approach has the great advantage of being easy to implement and insensitive to the confounders (such as smoking). The certainty is relatively high (95% confidence intervals are typically around $\pm 50\%$), but only part of the full impact is observed.

End-points that show up only after a longer period require observations of individuals or populations that are exposed to different levels of pollution. Dose–response functions for chronic effects are notoriously difficult to establish with confidence, and there are only few studies that have determined chronic CR functions. Of particular importance are the studies of Dockery *et al.*,[40] Pope *et al.*,[41] and Abbey *et al.*,[42,43] that find significant chronic effects of air pollution. The difference between chronic and acute CR functions is not so much in the exposure (most people are chronically exposed) as in the effects that are measured: do they show up within a few days after an exposure or only after a longer period? By analogy the terms acute and chronic are also applied to CR functions for mortality, even though the attributes appear strange in that context.

Particles

In air pollution studies particulate matter (PM, with subscript indicating largest aerodynamic diameter in μm that is included) designates anything that collects on the filters of particle detectors. It is a mixture of combustion particles, sulfate aerosols (including droplets of sulfuric acid) and nitrate aerosols, as well as particles from soil or sea spray. Most monitoring stations measure only the mass concentration of PM without any detail on composition. Unfortunately very little is known about the effects of individual components or characteristics (such as acidity, solubility, oxidizing potential, *etc.*) of PM.

As for size, particles of more than 10 μm diameter are stopped in the upper respiratory ducts and appear less harmful. Between 2.5 and 10 μm, the particles penetrate more deeply into bronchi and bronchioles; particles smaller than 2.5 μm reach the alveoli of the lungs. In the past most monitoring stations have measured PM_{10}; in recent years some have also measured $PM_{2.5}$. We assume a ratio of $PM_{2.5}/PM_{10} = 0.60$ based on typical ambient concentration data in the

[40] D. W. Dockery, C. A. Pope III, Xiping Xu, J. D. Spengler, J. H. Ware, M. E. Fay, B. G. Ferris and F. E. Speizer, An association between air pollution and mortality in six US cities, *New England J. Med.*, 1993, **329**, 1753–1759.

[41] C. A. Pope, M. J. Thun, M. M. Namboodri, D. W. Dockery, J. S. Evans, F. E. Speizer and C. W. Heath, Particulate air pollution as a predictor of mortality in a prospective study of US adults, *Am. J. Resp. Crit. Care Med.*, 1995,

[42] D. E. Abbey, M. D. Lebowitz, P. K. Mills, F. F. Petersen, L. W. Beeson and R. J. Burchette, Long-term ambient concentrations of particulates and oxidants and development of chronic disease in a cohort of nonsmoking California residents, *Inhalation Toxicol.*, 1995, **7**, 19–34.

[43] D. E. Abbey, N. Nishino, W. F. McDonnell, R. J. Burchette, S. F. Knutsen, L. W. Beeson and J. X. Yang, Long-term inhalable particles and other air pollutants related to mortality in nonsmokers, *Am. J. Resp. Crit. Care Med.*, 1999, **159**, 373–382.

USA and the EU. Particle emissions from modern incinerators are almost entirely PM_{10}.

Among the impacts quantified by ExternE, chronic mortality makes by far the largest contribution. That is based on three important cohort studies that have found effects of particulate air pollution on chronic mortality.[40,41,43] By far the largest of these, Pope *et al.*,[41] finds clear associations of mortality with fine particles ($PM_{2.5}$) and with sulfates. Here we use the $PM_{2.5}$ association of Pope *et al.*, a study whose analysis has recently been reconfirmed.[44] Since these chronic mortality studies determine a change in age-specific mortality, one can derive implicit estimates of the number of YOLL (years of life lost).[8,45,46]

Oxides of Nitrogen and of Sulfur

Whereas epidemiological studies in the US had generally concluded that direct effects of SO_2 did not appear significant, recent studies in Europe have found significant correlations for acute mortality and for respiratory hospital admissions; they have been used by ExternE.[8] In any case the resulting costs are relatively small. There have also been some studies that find direct effects of NO_x or NO_2, but ExternE concluded that they were not sufficiently convincing. NO_x is, however, implicated as precursor of nitrates and ozone.

Carbon Monoxide

Carbon monoxide (CO) is certainly toxic at concentrations much higher than found in typical urban environments. There seem to be harmful effects even at typical ambient concentrations, and several recent studies have proposed linear CR functions for CO. The evidence for a correlation with hospital admissions is quite strong and ExternE[8] has included it. There may also be mortality impacts due to CO but the case is less clear. The resulting damage cost is very small (see Figure 4, p. 191). One may well wonder if this is an artifact due to the inability of epidemiological studies to correctly identify the full impact of CO.

Ozone

The ExternE[8] estimate of the damage costs for ozone formation has been derived by Rabl & Eyre.[30] The step from cost per ppb ozone to cost per kg of NO_2 and per kg of VOC (volatile organic compound) precursor is based on results of the EMEP model for ozone formation[27] as well as the Harwell Global Ozone model.[47] The resulting cost is shown in Figure 4. Only a single European average of regional damages was derived.

44 HEI, *Particle Epidemiology Reanalysis Project*, Health Effects Institute, Cambridge MA, 2000. Available at http://www.healtheffects.org/pubs-recent.htm.

45 A. Rabl, Mortality risks of air pollution: the role of exposure–response functions, *J. Hazard. Mater.*, 1998, **61**, 91–98.

46 L. Leksell and A. Rabl, Air pollution and mortality: quantification and valuation of years of life lost. *Risk Anal.*, 2001, **21**, 843–857.

47 A. M. Hough, The development of a two-dimensional global tropospheric model. 2. Model chemistry, *J. Geophys. Res.*, 1991, **96**, 7325–7362.

Sulfate and Nitrate Aerosols

SO_2 and NO_x are transformed in the atmosphere to sulfates and nitrates, respectively, thus becoming a component of PM. ExternE[8] applies the CR functions for particles to these aerosols per concentration of pollutant mass. In particular, nitrate aerosols are considered like PM_{10} and sulfate aerosols like $PM_{2.5}$. There is much uncertainty about this. While there are studies that report correlations of mortality and other end-points with sulfates, there are no CR functions for nitrates because in the past nitrates have not even been monitored as a separate component of air pollution.

Other Pollutants

Among the heavy metals the following are considered carcinogenic: arsenic (As), cadmium (Cd), chromium (Cr, in oxidation state VI) and nickel (Ni). We assume the dose–response functions published by EPA.[48] Only inhalation dose has been taken into account because the available slope factors are for inhalation. By contrast to the classical air pollutants and carcinogens, for non-cancer impacts of heavy metals we only have data for thresholds below which no adverse effects have been observed. Therefore we have not yet quantified the impacts of Hg and Pb.

We also consider dioxin, a pollutant emitted from the incineration of municipal solid waste. The calculation is documented by Rabl, Spadaro and McGavran;[2] it includes all pathways, not just inhalation. However, for the present paper we base the dose–response function on EPA (2000)[3] rather than EPA (1994);[13,49] as a result our present dioxin impacts are an order of magnitude higher than the ones of 1998. We note that EPA (1994)[13,49] carries the label 'draft do not cite or quote' (as does EPA (2000)[3]), but no final version was released.

5 Monetary Valuation

The goal of the monetary valuation of damages is to account for all costs, market and non-market. For example, the valuation of an asthma attack should include not only the cost of the medical treatment but also the willingness-to-pay (WTP) to avoid any residual suffering. If the WTP for a non-market good has been determined correctly, it is like a price, consistent with prices paid for market goods. Economists have developed several tools for determining non-market costs; of these tools contingent valuation[50] has enjoyed increasing popularity in recent years. The results of well conducted studies are considered sufficiently reliable.

It turns out that damage costs of air pollution are dominated by non-market goods, especially the valuation of mortality. The single most important parameter is the 'value of statistical life' VSL (an unfortunate terminology for

[48] HEAST, *Health Effect Assessment Summary Table (HEAST)*, United States Environmental Protection Agency, Report EPA/540/R-95/036, Washington, DC 20460, May 1995.
[49] EPA, *Health Assessment Document for 2,3,7,8-Tetrachlorodibenzo-p-Dioxin (TCDD) and Related Compounds*, Report EPA/600/BP-92/001a, b and c, United States Environmental Protection Agency, Washington, DC 20460, June 1994.
[50] R. C. Mitchell and R. T. Carson, *Using Surveys to Value Public Goods: the Contingent Valuation Method. Resources for the Future*, Washington, DC, 1989.

what is really the 'willingness to pay for reducing the risk of an anonymous premature death'). In ExternE,[8] a European-wide value of 3.1 M€ was chosen for VSL, somewhat lower than similar studies in the USA; this value was chosen as average of the VSL studies that had been carried out in Europe. The uncertainty is large and one could argue for other values in the range of 1–5 M€.

A crucial question for air pollution mortality is whether one should simply multiply the number of premature deaths by VSL, or whether one should take into account the years of life lost (YOLL) per death. The difference is very important because premature deaths from air pollution tend to involve far fewer YOLL per death than accidents (on which VSL is based). The ExternE[8] numbers, used here, are based on YOLL and thus significantly lower (for the same dose–response function) than the simple VSL valuation assumed in most previous external cost studies. There is considerable uncertainty about the relation between VSL (which has been determined for accidents) and the value of a YOLL due to air pollution, because it involves the period at the end of life about which valuation studies are only just beginning. In ExternE[8] the value of a YOLL has therefore been calculated on theoretical grounds by considering VSL as the net present value of a series of discounted annual values. The ratio of VSL and the value of a YOLL thus obtained depends on the discount rate; it is typically in the range of 20–40.

For the value of a YOLL ExternE[8] assumes 0.083 M€ for chronic mortality, and 0.155 M€ for acute mortality, the difference arising from assumptions about latency and discounting. Due to discounting the value of a YOLL for chronic mortality is lower than for acute mortality because of the time delay between exposure to a pollutant and the premature death. We also assume 0.45 M€ for non-fatal cancers and 1.5 M€ for fatal cancers.

6 Calculation of Damage

Site Dependence of Impacts

The total damage D due to a quantity Q of a pollutant is obtained by integrating the damage at a point x over all points x of the region affected by the pollutant; the damage at a point x is the product of the receptor density $\rho(x)$, the slope of s_{CR} of the CR function, and the concentration increment $c(x)$ due to Q.

The damage depends on the site where the pollutants are emitted. Site dependence is illustrated in Figure 3 for primary pollutants, *i.e.* pollutants harmful in the form emitted by a source. This figure shows two variations at once: with stack height, and with source location for five specific sites in France. Plume rise is included for typical conditions of large combustion installations. As an example, we consider a specific impact: the increase in acute mortality (YOLL) due to an emission of $Q = 10^6$ kg yr^{-1} of SO_2 (chosen arbitrarily). The damage is shown on two scales, as number of YOLL per year on the right hand scale, and in units of D_{uni} (to be explained in the following section) on the left. At a stack height of 100 m the impact for the site near Paris is about 3 times larger than D_{uni} and for Cordemais (a relatively rural site on the Atlantic Ocean) it is only 0.4 times D_{uni}. The impact for Martigues is rather small, despite the proximity of a large city,

Figure 3 An example of dependence on site and on height of source for a primary pollutant with linear dose-response function: damage D from SO_2 emissions, for five sites in France, in units of D_{uni} for uniform world model Eq. (3) (the nearest big city, 25–50 km away, is indicated in parentheses). Scale on right indicates $YOLL\ yr^{-1}$ (acute mortality) from a plant with emission $10^6\ kg\ yr^{-1}$. From Rabl and Spadaro,[4] with permission from the *Annual Review of Energy and the Environment*, Vol. 25, © 2000 by Annual Reviews

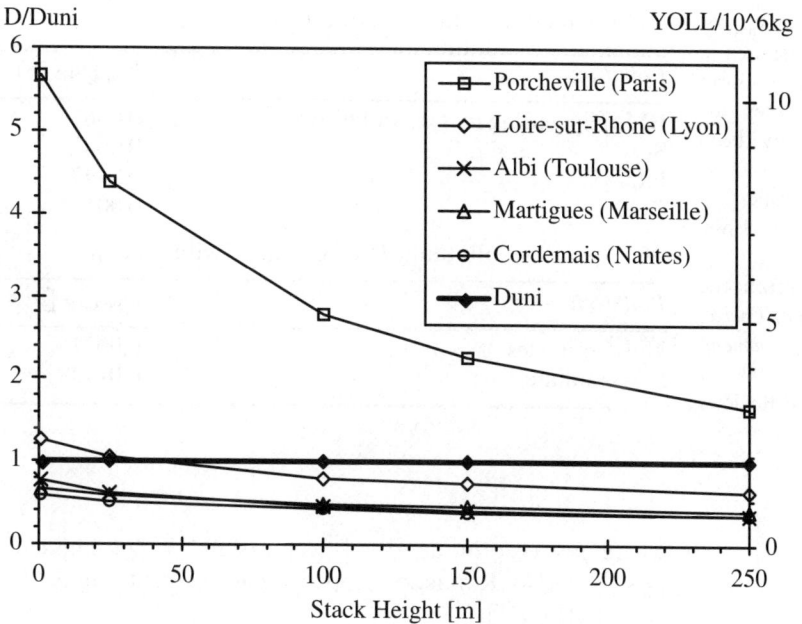

because the prevailing wind carries the pollutants out to sea.

For secondary pollutants (created by chemical reactions of primary pollutants) such as sulfates, nitrates and ozone, the sensitivity to local detail is much lower because these pollutants are not created until some distance from the source. For nitrates and sulfates this occurs over tens of km from the source, and so the site dependence is quite weak. Based on EcoSense results, we estimate that variations of sulfate or nitrate damage, per kg of SO_2 or NO_2, with site are around 50% in Central Europe. The creation of ozone is more rapid, within several km to tens of km from the source, and there are strong regional variations in the chemistry; based on EMEP data we estimate that ozone damage per kg of precursor could vary with site by about a factor of four in Europe.[30]

A Simple Model for Typical Damage Estimates

Curtiss and Rabl[51] have shown that the total damage D for a linear dose–response function and stationary conditions can be calculated in closed form if one assumes that the receptor density and the depletion velocity of the pollutant from the atmosphere are the same everywhere in the affected region. The depletion velocity is defined as ratio the of the total depletion flux (due to dry deposition, wet deposition or chemical reactions) and the concentration. Under these conditions of uniformity one finds, as a consequence of mass conservation, the following very simple 'uniform world model' for the total damage rate D_{uni} (in impact units of the CR function)

[51] P.S. Curtiss and A. Rabl Impacts of air pollution: general relationships and site dependence, *Atmos. Environ.*, 1996, **30**, 3331–3347.

Table 4 Values of k_{uni} for different pollutants, used in this paper for 'uniform world model'. Except for CO, they have been derived by non-linear regression to dispersion calculations with EcoSense. From Rabl and Spadaro,[4] with permission from the *Annual Review of Energy and the Environment*, Vol. 25, © 2000 by Annual Reviews

(a) Primary pollutants, D_{uni} of Eq. (3) with k_{uni}

Pollutant	k_{uni} $[m\ s^{-1}]$
PM_{10} (including metals and dioxins)	0.0067
SO_2	0.0073
NO_2	0.0147
CO	0.001

(b) Secondary pollutants, D_{2uni} of Eq. (5) with $k_{2uni,eff}$

Pollutant	$k_{2uni,eff}$ $[m\ s^{-1}]$
$NO_2 \rightarrow$ nitrates	0.0071
$SO_2 \rightarrow$ sulfates	0.0173

$$D_{uni} = s_{CR}\rho Q/k \tag{3}$$

where ρ = receptor density [receptors m^{-2}], Q = emission rate of pollutant [g s^{-1}], s_{CR} = CR function slope [impacts/(receptor·(g m^{-3}))] and k = depletion velocity [m s^{-1}].

The quantity Q/k represents the concentration increment due to Q, averaged over the affected receptors. As an illustration we calculate D_{uni} for the CR function used in Figure 3 whose slope is $s_{CR} = 5.34E - 06$ YOLL/(pers yr μg m^{-3}). The depletion velocity is $k = 0.0073$ m s^{-1} for SO_2 (see Table 4). Inserting these numbers into Eq. (3) with the regional average population density $\rho = 8.0E - 5$ person m^{-2} (see below) we obtain for $Q = 10^6$ kg yr^{-1} = 3.17E07 μg/s^{-1}:

$$D_{uni} = \frac{5.34 \times 10^{-6} \text{YOLL/(person yr } \mu g\ m^{-3}) \times 8.0 \times 10^{-5} \text{ person m}^{-2}}{0.0073\ m\ s^{-1}}$$
$$\times 3.17 \times 10^7\ \mu g\ s^{-1} = 1.86 \text{ YOLL yr}^{-1} \tag{4}$$

This is shown as the thick horizontal line in Figure 3. It lies right in the middle of the curves for the five sites.

Eq. (3) can be generalized to yield the damage D_2 due to a secondary pollutant:

$$D_{2uni} = \frac{s_{CR2}\rho_{uni}}{k_{2uni,eff}} Q_1 \tag{5}$$

where Q_1 = emission rate of the primary pollutant, s_{CR2} = CR function slope of the secondary pollutant and $k_{2uni,eff}$ = 'effective depletion velocity', taking into account the depletion velocities of the primary and secondary pollutants and the transformation rate.

We have determined numerical values of k_{uni} and $k_{2uni,eff}$ by fits to the dispersion results of EcoSense. Table 4 shows the values of k_{uni} that we will use in this paper. For carbon monoxide we assume 0.001 m s^{-1} based on life time estimates of 36–110 days given by Manahan,[52] about a factor ten longer than for PM_{10} and SO_2.

[52] S. E. Manahan, *Environmental Chemistry*, CRC Press, Boca Raton, FL 33431, 1994.

In the region bounded by Sicily to the South, Portugal to the West, Scotland to the North and Poland to the East, the average population density is 80 persons km^{-2}. This is about half the average EU15 population density of 158 persons km^{-2} per land area because it includes much water. Examining the results of detailed site specific calculations for more than fifty installations in the EU15 countries,[53] we have found that D_{uni} of Eqs. (3) and (5), with k_{uni} of Table 4 and $\rho_{uni} = 80$ persons km^{-2}, does indeed provide representative results for stack heights of 50–100 m.

The reason for this remarkable success of the simple 'uniform world model' is that with tall stacks much of the total impact occurs in regions sufficiently far from the source where the pollutant is fairly well mixed vertically in the planetary boundary layer, and variations of the depletion velocity are not too large. Furthermore, averaging site-specific results over a range of emission sites is mathematically equivalent to averaging over population distributions, thus bringing the effective population distribution closer to uniformity. If one wants typical results for public policy, without being able to evaluate each and every site, D_{uni} seems as good a choice as any – and it has the advantage of being simple and transparent. Also, it is a convenient tool for estimating damages for sites outside the EU. Therefore we will use D_{uni} for all the damage estimates in this paper.

7 Damage Costs per Kilogram of Pollutant

The key assumptions for our calculations are summarized in Table 5. In Figure 4 we present the results for the damage costs per kg of pollutant, obtained with the 'uniform world model' with the parameters of Section 4, in particular a population density of $\rho = 80$ persons km^{-2}. For other regions these numbers should be scaled according to regional average population density (land and water, within a radius of 500–1000 km). The multipliers in the caption of Figure 4 indicate how much the numbers could change for different conditions. The error bars in Figure 4 indicate the uncertainties.[29]

8 Damage Costs per Kilogram of Waste

Scope of Analysis

It has become fashionable to carry out life cycle assessments (LCA), taking into account all stages of a process from cradle to grave. However, the appropriate system boundaries for an LCA depend on the question under consideration, a point often overlooked. For example, a cost–benefit analysis of flue gas clean up technologies for incinerators should take into account only the incineration, not other stages of waste management. A comparison of incineration with landfill should take into account all stages downstream of waste collection; emissions from transport of waste are only relevant to the extent that transport modes and distances might differ between landfill and incineration. Avoided emissions due

---

[53] J. V. Spadaro, *Quantifying the Effects of Airborne Pollution: Impact Models, Sensitivity Analyses and Applications*, Doctoral thesis, Ecole des Mines, 60 boul. St.-Michel, F-75272 Paris, France, 1999.

Table 5 Key assumptions for the calculations in this paper. From Rabl and Spadaro,[4] with permission from the *Annual Review of Energy and the Environment*, Vol. 25, © 2000 by Annual Reviews

Atmospheric dispersion models	
Local range:	ISC gaussian plume model[24]
Regional range (Europe):	Harwell Trajectory Model as implemented in EcoSense software of ExternE. Ozone impacts based on EMEP model,[27] as interpreted by Rabl and Eyre.[30]
Impacts on health	
Form of dose–response functions	Linearity of incremental impact due an incremental dose for all health impacts
Chronic mortality	CR function slope $s_{CR} = 4.1E-4$ YOLL (years of life lost) per person per year per $\mu g\,m^{-3}$ derived from increase in all-cause age-specific mortality due to $PM_{2.5}$,[41] by integrating over age distribution and assuming it applies only to people over age 30
Acute mortality	For SO_2 and ozone, with 0.75 YOLL per death
Nitrate and sulfate aerosols	CR functions for nitrates same as for PM_{10} CR functions for sulfates same as for $PM_{2.5}$ (slope = 1.7 times slope of PM_{10} functions)
Micropollutants	Only cancers have been quantified (As, Cd, Cr, Ni and dioxins); effects of Hg and Pb have not been quantified
Monetary valuation	
Valuation of premature death	Proportional to reduction of life expectancy, with value of a YOLL (years of life lost) derived from VSL = 3.1 M€: $v_{YOLL} = 0.083$ M€ for chronic mortality $v_{YOLL} = 0.155$ M€ for acute mortality
Valuation of cancers	0.45 M€ non-fatal cancers 1.5 M€ for fatal cancers

to energy recovery (Section 2.3) are relevant for a comparison of incineration with other waste treatment options, or for a decision whether to locate an incinerator close to a district heating system.

Impacts downstream of an incinerator arise from the residues. They are relevant for a comparison of waste treatment options, but not for an evaluation of flue gas clean up (except to the extent that flue gas clean up may affect the residues). In passing we note that downstream impacts are uncertain to the extent that they depend on the future management of the residues.

In this paper we do not consider upstream or downstream impacts, drawing the boundary at the incinerator instead. Of course, for other system boundaries the results can readily be integrated in the appropriate LCA.

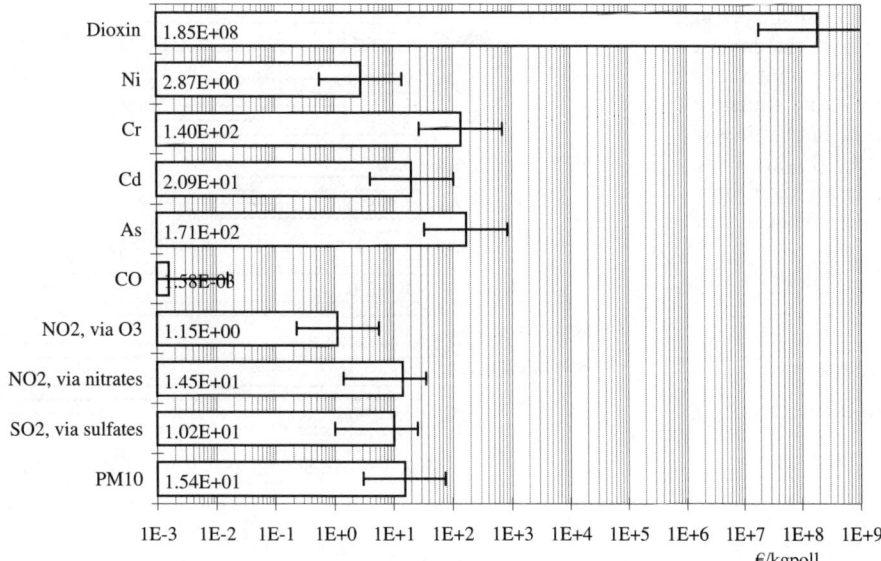

Figure 4 Mean damages per kg of pollutant emitted by incinerators in Europe. Error bars indicate uncertainties as 1 geometric standard deviation intervals

Multipliers for **variation with site** (proximity of big city, local climatic conditions) and **stack conditions** (stack height h, temperature T, exhaust velocity v): No variation for CO_2; Weak variation for dioxin because non-inhalation pathways dominate: $\approx 0.7–1.5$; Weak variation for secondary pollutants: $\approx 0.5–2.0$; Strong variation for primary pollutants: $\approx 0.5–6$ for site, $\approx 0.6–3$ for stack conditions (up to 15 for ground level emissions in big city).[53]

Results

Figure 5 shows the implications for the damage per kg of waste, obtained by multiplying the $€\,kg^{-1}$ numbers of Figure 4 by the emissions per kg of waste, second column of Table 2. Despite their high toxicity, the micropollutants do not dominate the total damage. This is especially striking for dioxins whose toxicity is extreme but whose emission rate is so low that the damage per kg of waste is small compared to the classical air pollutants. Damage of NO_x *via* nitrates appears to dominate, but we recall the comment in Section 4.5 about the uncertainty of this pathway. Note that the damage estimates for the classical air pollutants are expectation values, whereas the ones for micropollutants are upper bounds (95% confidence) because of the definition of the slope factor.

The sum of the health damage costs per ton waste is approximately 20 € per 1000 kg. It is interesting to note that the market cost of MSW incineration in France was around 70 € per 1000 kg with the old regulations and was expected to increase about 25% with the new regulations.[54] Therefore any remaining damage costs after imposition of the EC 2000 Directive are relatively small compared to the market cost of MSW incineration.

[54] J.-P. Peyrelongue, GEC ALSTHOM Stein Industrie, Variations et différences de référentiels: influence sur les coûts (Variations and different references: influence on the costs), Présentation EUROFORUM, Paris. 3 juin 1997.

Figure 5 Damage per kg of waste, for the damage per kg of pollutant values in Figure 4 and emissions equal to limit values of EC[1] (Figure 1 and Table 1): (a) logarithmic scale; (b) linear scale

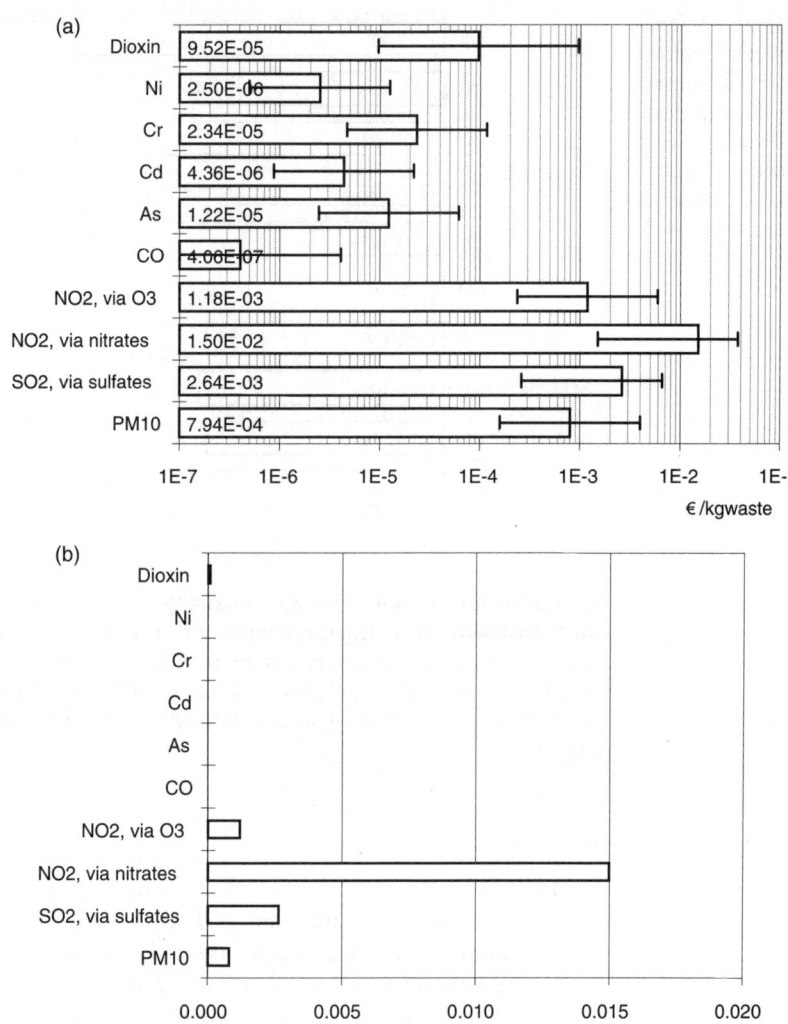

9 Conclusions

We have assembled data on emissions, ambient concentrations and damage costs for air pollutants from incineration of waste. Issues of site dependence and geographic extent of the damage have also been addressed. For the quantification of damages and costs we have assumed linear dose–response functions.

On the basis of our data and calculations we have presented several comparisons to put the risks of incinerators in perspective. Assuming $500 \, kg \, yr^{-1}$ as typical per capita production of MSW, incinerated with emissions equal to the limit values of EC[1] we have made several comparisons, to the extent that we have been able to find suitable data:

- incremental concentration compared to background concentration (Section 3);
- incremental concentration (or dose) compared to health guidelines (Section 3);
- health risks of different pollutants compared to each other (Section 8);
- incremental damage cost compared to the cost of incineration itself (Section 8).

Rabl, Spadaro and McGavran[2] provide two additional comparisons, based on the same assumptions,

- incremental emission compared to other emissions (natural and anthropogenic, *e.g.* cars);
- incremental years of life lost compared to other risks of everyday life.

By any of these comparisons the health impacts of MSW incineration appear insignificant, if the emissions respect the EC regulations.[1]

In particular, the incremental concentration of pollutants from such incinerators is far below the ambient air quality guidelines of EPA[33] or WHO.[32] Since current background concentrations are usually also well below such guidelines, so is the total. Thus there would be no damage *if* these guidelines were no-effect thresholds. That does not guarantee the complete absence of harmful effects, but, whatever they may be, air quality guidelines do not suffice to quantify them.

The cancer impacts of micropollutants, in particular of dioxins, are small compared to the mortality due to ordinary particulate matter from MSW incinerators which in turn is insignificant compared to the contribution of other sources of particulate matter or compared to other risks of everyday life. Similar conclusions about dioxins have also been reached by Eduljee and Gair.[55]

An assessment of the impacts of MSW incineration must not overlook the benefits of avoided emissions from energy recovery, for district heating, industrial process heat or power generation: the net emissions can be quite small or even negative. Even though MSW is a dirty fuel, the clean technologies required by EC[1] or equivalent regulations ensure that the emissions will be lower than from most conventional combustion technologies with much cleaner fuels.

Finally we note that regulations alone do not guarantee good environmental performance. Verification is crucial. As a good example we might mention the incinerator of the City of Vienna which publicly displays measured emissions data in real time.[19]

Acknowledgements

This work has been supported in part by the ExternE Project of the JOULE Program and by the Environment and Climate Program of the European Commission DG Research, by ADEME, and by the Ministère de l'Environnement. We thank Peter Curtiss, Mark Delucchi, William Dab, Brigitte Desaigues, Nick Eyre, Rainer Friedrich, Mike Holland, Fintan Hurley, Wolfram Krewitt, Pat McGavran, Jean-Pierre Tabet and Philippe Tiberghien for valuable discussions.

[55] G.H. Eduljee and A.J. Gair, Setting the dioxin limits for MSW incinerators: a multimedia exposure assessment framework, *Waste Manage. Res.*, 1997, **15**, 335–348.

Methodological Issues Related to Epidemiological Assessment of Health Risks of Waste Management

HELEN DOLK

1 Introduction

Waste management is an essential sector of human activity. Should we bury, burn or recycle it, or should we produce less of it? What is the appropriate mix of these four options? What technology should we employ for containment or treatment, how far should we transport wastes, where should we site waste disposal and treatment centres, which processes should we target for waste reduction? There is clearly a balance to be made between the health and wellbeing benefits of production and the health and wellbeing costs of waste. Rational decision making about waste management should be based on optimisation of health and wellbeing.

It is relevant also to ask when considering alternative waste management strategies who is experiencing the health and wellbeing benefits of production, and who is bearing the health and wellbeing costs of waste? Are these benefits and costs distributed evenly across the population, within and between countries? How do our different waste management options increase or reduce inequalities?

Results of epidemiological studies of the health risks of waste management can be useful 'bricks' in health impact assessment,[1] but the limitations of these studies are as striking as their potential to enlighten. Traditional epidemiology is essentially a reductionist activity, focusing on individual cause–effect relationships. Epidemiology is comfortable looking at the health effects of smoking, but struggles when looking at the health effects of complex issues such as waste management or global warming,[2] which requires recourse to modelling. For example, adverse health impacts of landfill as a component of waste management

---

[1] K. Lock, Health Impact Assessment, *Br. Med. J.*, 2000, **320**, 1395–1398.

[2] A. J. McMichael, The health of persons, populations and planets: epidemiology comes full circle, *Epidemiology*, 1995, **6**, 633–636.

Issues in Environmental Science and Technology, No. 18
Environmental and Health Impact of Solid Waste Management Activities
© The Royal Society of Chemistry, 2002

strategy potentially include effects of exposure to toxic chemicals through emissions to air, water or soil, effects of exposure to infection and biological contaminants, stress related to odour, noise, vermin, visual amenity or local economic changes, risk of accidents such as fire, explosions or subsidence, transport-related health impacts including accidents, spills, air pollution and energy consumption, and health impacts of global warming from landfill methane and transport emissions. Set against this are the health benefits of the waste creating activities, of the potential for energy recovery from landfills and of the economic activity in the landfill sector. All of this must also be assessed in terms of inequalities as mentioned above.

I am going to discuss in this paper some methodological issues in epidemiological approaches to the assessment of health risks of waste management, accepting that what I will be focusing on is a few individual bricks, rather than the whole building of health impact assessment and accepting that the final design of the building of health impact assessment may not be clear or agreed. I do not attempt to review the epidemiological literature, which is done elsewhere in this volume and elsewhere.[3-5] For illustrative purposes, I refer particularly to the EURO-HAZCON study[6] which investigated the relationship between living near hazardous waste landfill sites and risk of congenital anomalies in Europe, since I am particularly familiar with its scientific methodology and limitations, and with how it has been viewed by different stakeholders.

2 The Design of an Epidemiological Study

Epidemiology is the study of the distribution and determinants of disease in the population. 'Disease', in the context of health impact assessment of waste management, is a convenient shorthand for suboptimal health and wellbeing, in accordance with the WHO definition of Health as 'physical, mental and social well-being'. It is central to epidemiologic design that we compare groups of people. These groups may differ in one or more relevant 'risk factors'. The classic epidemiological question is to ask whether factor F leads to, or has some part to play in leading to, disease D. Epidemiology highlights why different people have different diseases.

The ideal way of answering whether factor F leads to disease D is to organise an experiment whereby all factors other than F are held constant between the two groups. Although randomised controlled trials are important in branches of epidemiology seeking to evaluate health service interventions, they are rarely practical in the field of environmental epidemiology which must rely on observational studies instead. The interpretation of observational studies is

[3] B.L. Johnson, *Impact of Hazardous Waste on Human Health*, CRC Press, 1999.
[4] National Research Council, *Environmental Epidemiology*, Vol 1, *Public Health and Hazardous Wastes*, National Academy Press, Washington, DC, 1991.
[5] M. Vrijheid, Health effects of residence near hazardous waste landfill sites: a review of epidemiologic literature, *Environmental Health Perspect.*, 2000, **108** (suppl 1), 101–112.
[6] H. Dolk, M. Vrijheid, B. Armstrong, L. Abramsky, F. Bianchi, E. Garne, V. Nelen, E. Robert, J.E.S. Scott, D. Stone, R. Tenconi, Risk of congenital anomalies near hazardous waste landfill sites in Europe: the EUROHAZCON study, *Lancet*, 1998, **352**, 423–427.

centrally concerned with assessing the potential for bias, usually from unrecognised differences between the groups being compared. These differences may be in the way that disease has been measured, in the way that exposure has been measured, or in the presence of confounding factors.

There are a number of main designs for an observational epidemiological study. Case-control designs select a group of 'cases' with disease D, and a group of 'controls' without disease D, and then set about determining the presence or strength of a predetermined set of risk factors in each group. The question is whether a greater proportion of cases than controls have a certain risk factor present. Cohort designs identify an 'exposed cohort' where a risk factor F is present, and a control cohort where risk factor F is absent, and then set about determining the presence or severity of a predetermined set of health or disease outcomes in each cohort. The question is whether a greater proportion of people with a certain risk factor/exposure develop disease D than without that risk factor/exposure. An Ecological study considers population subgroups rather than individuals as its units of observation. In each population subgroup, the frequency of one or more risk factors is measured, as well as the frequency of one or more diseases. The question then is whether population subgroups which have higher levels of a particular risk factor also have higher levels of a particular disease.

Epidemiological surveillance differs in its emphasis from analytic epidemiology. Surveillance emphasises continous monitoring using rather limited routine data collection designed to address a range of general hypotheses. One of the objectives of the EUROHAZCON study,[6] for example, was to test the feasibility of using a pre-existing European congenital anomaly surveillance network (EUROCAT) for environmental health surveillance. The term 'envirovigilance' would be appropriate to describe environmental health surveillance in analogy with pharmacovigilance or post-marketing surveillance of adverse effects of therapeutic drugs. Such surveillance is based on recognition of the limitations of pre-licensing/marketing risk assessment. Development of envirovigilance is particularly needed since the majority of industrial chemicals and their break-down products have not been subjected to rigorous toxicity and teratogenicity testing, and because industrial activities, including waste management, do not necessarily follow the best practice guidelines upon which predictive risk assessments are based. Surveillance can be useful to direct attention to areas where more analytic epidemiological studies are needed.

3 Disease/Health Measurement

Health status measurement may range from death and life threatening diseases requiring medical treatment to quality of life measures. Landfills and incinerators, for example, may impact on health through biological effects of toxic chemical exposures, or through stress associated with smell, visual, noise, local economic or other aspects of waste site operation. Stress may also occur because of fear of chemical toxicity. Understanding the pathways to ill health is important for preventive measures. In terms of epidemiological study however, the first task is simply to determine whether there is a difference in health/disease between people with different exposures related to waste management.

Other than observing whether someone is dead or alive, most health/disease measurements are subject to some level of observer variation in diagnosis, assessment or reporting. This cannot be ironed out for the convenience of epidemiology. The important question is whether health/disease are being measured consistently in exposed and unexposed people, or whether for any reason exposure status may affect the likelihood that people of equal health status are diagnosed as having a disease or not. If either the researcher, health assessor or the subject of the study knows that disease may be related to exposure at the time the health assessment is made, then they may be influenced in their assessment leading to bias in the results. For example, a study may send a questionnaire to all members of a community to study the health impact of an incinerator or landfill. If the questionnaire concerns health assessments on the more 'subjective' end of the spectrum, for example headaches and feelings of nausea, people who perceive themselves as exposed (whether because they live nearby or suffer more smells from the facility) may be more likely to report these symptoms, as has been shown in a number of studies of landfills.[5] Furthermore, people concerned by the possible effect of the exposure on their health may be more likely to return the questionnaire than others, particularly if they feel their health has been impaired. However, it is important to be very clear about exactly what research questions the study is asking, and what issues need to be addressed beyond these questions. If people perceiving themselves to be exposed to the waste facility self-rate their health more poorly than others, this may have a real adverse impact on their quality of life which needs to be addressed even when direct chemical toxicity is not the underlying mechanism.

Cancer and congenital anomalies are generally on the less subjective end of the disease spectrum. However, it still matters how information is gathered. For example, a community group or local health authority may survey a community close to a waste facility, ascertaining cases of cancer or congenital anomaly in the population. If they compare this to routinely collected health statistics, they may well find a different rate in the local community, due to differences in the completeness of case finding, or inclusion or exclusion criteria. It is important therefore for studies to use the same source of information to measure disease in both exposed and unexposed populations. Studies based on disease registers, where cases are ascertained blind to the study hypothesis, have an advantage in this regard.

Some studies have involved surveys of the population with direct measurement of physiological status or markers of biological effect. For example, levels of liver enzymes were measured in a community whose drinking water was contaminated by chlorinated compounds from a landfill site,[7] some having used more of this contaminated water than others. A study of the Mellery landfill site in Belgium looked at chromosomal changes.[8,9] Such measurements are free of reporting bias

---

[7] C.S. Clark, C.R. Meyer, P.S. Gartside, V.A. Majeti, B. Specker, W.F. Balisteri and V.J. Elia, An environmental health survey of drinking water contamination by leachate from a pesticide waste dump in Hardeman County, Tennessee, *Arch. Environ. Health*, 1997, **37**, 299–310.

[8] T. Lakhanisky, D. Bazzoni, P. Jadot, I. Joris, C. Laurent, M. Ottogali, A. Pays, C. Planard, Y. Ros, C. Vleminckx, Cytogenetic monitoring of a village population potentially exposed to a low level of environmental pollutants. Phase 1: SCE analysis, *Mutat. Res.*, 1993, **319**, 317–323.

Figure 1 Hierarchy of exposure data or surrogates (source: ref. 4)

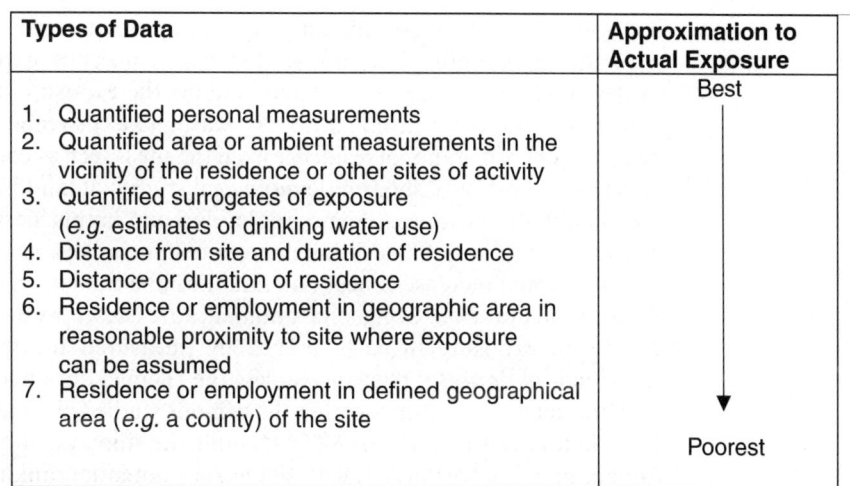

Types of Data	Approximation to Actual Exposure
1. Quantified personal measurements 2. Quantified area or ambient measurements in the vicinity of the residence or other sites of activity 3. Quantified surrogates of exposure (*e.g.* estimates of drinking water use) 4. Distance from site and duration of residence 5. Distance or duration of residence 6. Residence or employment in geographic area in reasonable proximity to site where exposure can be assumed 7. Residence or employment in defined geographical area (*e.g.* a county) of the site	Best Poorest

and laboratory analysis can be done blind to exposure status. An interesting study in Flanders[10] recruited adolescents living in the area of two incinerators, a lead smelter and a control rural area, measuring glomerular or tubular renal dysfunction, DNA damage and testicular volume. It is not always clear how biological markers relate to risk of disease, but at least such studies clarify the presence of a biological effect. It then remains to identify the exposures or confounders concerned.

We may sometimes tend to 'search under the lamp-post' for adverse health outcomes – choosing those which are easiest to collect data on or have been suggested in previous studies. An area which may be understudied, for example, is immunotoxicity and immune status.[3]

4 Exposure measurement

A hierarchy of exposure assessment has been proposed, which is reproduced in Figure 1.[4] At the top of the hierarchy are quantified personal measurements, including personal monitoring of contaminants in the breathing zone of the individual, and biological markers which integrate dose from multiple routes of exposure (*e.g.* urine analysis may indicate dose from the combination of ingestion of contaminants in food and water, inhalation and dermal exposures). Second in the hierarchy are area or ambient measures of exposure which are not personal

[9] W. Klemans, C. Vleminckx, L. Schriewer, I. Joris, N. Lijsen, A. Maes, M. Ottogali, A. Pays, C. Planard, G. Gigaux *et al.*, Cytogenetic biomonitoring of a population of children allegedly exposed to environmental pollutants. Phase 2: Results of a three year longitudinal study, *Mutat Res.*, 1995, **342**, 147–156.

[10] J. A. Staessen, T. Nawrot, E. Den Hond, L. Thijs, R. Fagard, K. Hoppenbrouwers, G. Koppen, V. Nelen, G. Schoeters, D. Vanderscheuren, E. Van Hecke, L. Verschaeve, R. Vlietinck and H. A. Roels, for the Environment and Health Study Group, Renal function, cytogenetic measurements, and sexual development in adolescents in relation to environmental pollutants: a feasibility study of biomarkers, *Lancet*, 2001, **357**, 1660–1669.

measures, but presumably reflect average personal exposure. At the bottom of the hierarchy (in seventh place) are studies which measure exposure in terms of residence within a large area which contains the exposure (the landfill site or incinerator). One up from there is 'reasonable proximity' to a site, and measurement of distance of residence from the site is seen as conferring additional advantage. Most published epidemiological studies of landfills and incinerators have a level of exposure assessment at levels 5 or 6 in this hierarchy. Duration of exposure can be a useful additional measure if it is relevant. Longer duration of exposure would increase cancer risk and thus one can classify exposure by both distance and duration. Congenital anomalies are caused by events in a short time window of exposure in early pregnancy (measured in weeks). Duration of exposure is of doubtful relevance unless the chemical has a long biological half life, although it is relevant whether the persons studied were resident during that time window. In the EUROHAZCON multisite study we attempted to combine distance of residence (level 5) with site hazard potential ranking thus combining both 'within site' and 'between site' exposure classifications.[11]

Currently in Europe, 'between site' exposure classifications are difficult. For example, a government report in UK stated in 1992 'In the UK it is not possible to characterise the majority of landfills even to the level where a simple risk assessment framework can be employed on a site-specific basis. This particularly applies to the characterisation of the emplaced waste'.[12] Waste management facilities with the most monitoring data may be precisely those where there is least cause for concern, as old, disused or badly managed or regulated facilities may lead to the most significant exposures. In the EUROHAZCON study, hazard potential ranking was done according to characteristics such as management, hydrogeology, engineering characteristics, and type of waste dumped.[11] Although a good level of consensus was achieved by a panel of landfill engineers carrying out the ranking, the panel were concerned about the extent of missing information and it was difficult to be confident that the final ranking was a true reflection of 'hazard potential'.

Studies based on distance of residence are often cheaper for a given sample size, as they can largely use routinely collected data. Again it is important to distinguish exactly what question is being asked. If residents near a landfill or incinerator are concerned that their health is poorer than those further away, then it may be legitimate to assess the epidemiological evidence for this, whether or not there is also evidence that they are more exposed to substances or processes of potential concern. The problems arise when using distance of residence as a proxy for general exposure. Firstly, by assessing the health of local residents, whatever answer is obtained, one must remember that this does not encompass the entire potential health impact, as discussed in the Introduction. Even local contamination may have distributed effects, such as wider population consumption of groundwater contaminated by landfill leachates, or entry into

[11] M. Vrijheid, H. Dolk, B. Armstrong, G. Boschi, A. Busby, T. Jorgensen, P. Pointer and the EUROHAZCON collaborative group. Hazard potential ranking of waste landfill sites and risk of congenital anomalies, *Occup. Environ. Med.*, 2002.

[12] *The technical aspects of controlled waste management. Health effects from hazardous waste landfills site*, Department of Environment Waste Technical Division, Report No. CWM/057/92.

the food chain and distributed consumption of dioxin from incinerator emissions. Secondly, working low in the exposure assessment hierarchy provides relatively weak evidence of causation relating to the impact of the landfill/incinerator on health. Such a crude measure of exposure tends to underestimate any real risks due to misclassification of the more exposed as less or unexposed and *vice versa* and may leave more opportunity for confounding.

Working with distance of residence only, there are many reasons for misclassification of relative degree of true exposure, including the fact that exposure is not likely to radiate evenly in all directions from the landfill, the fact that people vary in their lifestyles and resulting probability of exposure (*e.g.* do they stay at home or work elsewhere?, do they engage in recreational activities nearer or further from the landfill?, do they garden? *etc.*), and the fact that people often move house between exposure and disease diagnosis, and thus some of those considered 'exposed' did not live in the area at the relevant time, and *vice versa*.

Wind direction is a commonly quoted objection to studies using distance of residence as a surrogate of exposure. Surely those in the path of the prevailing wind would be more exposed, and further from the landfill, and the 'circle' of exposure should be replaced by a more directional ellipse? This is an area where more exposure assessment research is needed. Measurements taken near cokeworks have shown that air pollution on still days is the heaviest, regardless of the path of the prevailing wind.[13] The exposure model needed may vary depending on the health outcome considered. For congenital anomalies, it may be peak exposures in short time windows of fetal development that matter, and thus it may be more relevant to consider maximum exposures on still days close to the source. For cancer, it may be more relevant to average exposures over long time periods on a stochastic model, and thus wind direction may be more relevant.

Another problem in exposure classification is the not unusual situation of people living near to more than one waste disposal site. A British study suggested that 80% of the population in England live near (within 2 km) open or closed landfill sites, and often near more than one.[14] Logically, living near to two sites confers more exposure than living near one site, if indeed there is any exposure of relevance at all. But to classify exposure for people living 1 km from a single site compared to 2 km from two different sites, or 3 km from one site and 0.5 km from another, implies some model of how exposure attenuates with distance. At present, such a model is not available and moreover is likely to vary from site to site. The EUROHAZCON study and others have therefore used simple algorithms such as 'distance to nearest site', while acknowledging the extra potential for misclassification that this introduces.

Misclassification of exposure or poor exposure classification is often cited as a reason to disregard the results of studies such as EUROHAZCON which find evidence of an excess of disease near a waste disposal or treatment site. However, misclassification of exposure generally leads to underestimation, not overestimation,

---

[13] R. S. Bhopal, P. Phillimore, S. Moffat *et al.*, Is living near a coking works harmful to health?, *J. Epidemiol. Comm. Health*, 1994, **48**, 237–47.

[14] P. Elliott, D. Briggs, S. Morris, C. de Hoogh, C. Hurt, T. K. Jensen *et al.*, Risk of adverse birth outcomes in populations living near landfill sites, *Br. Med. J.*, 2001, **323**, 363–368.

of relative risks. This is quite intuitive – if the distinction between the exposed and unexposed is so poor that it is almost random, any real problem among the true exposed would be unlikely to be detected. Misclassification does not make observation of a spurious excess risk where there is in fact none more likely. Thus, this argument would lead us to conclude that the real risks may be higher than those revealed by the study.

If it were established beyond doubt that relevant exposure to landfills is entirely uncorrelated with distance of residence from the landfill, then it also follows that any observed excess among nearby residents cannot be causally related to the landfill. This, however, leaves us entirely back at square one in terms of determining the health effects of landfill, with the need to choose an exposure measure which does correlate with relevant aspects of 'true' exposure.

Misclassification of exposure by using a crude distance measure can theoretically lead to some seemingly anomalous results. For example, if one is studying a single site and wind direction is important to exposure, then it could be that patterns of settlement mean that those living near the site (*e.g* within 1 km) happen to be to the east out of the path of the wind, and those further from the site (*e.g.* 1–2 km) are mainly in the west in the path of the wind. The observed relative risks would then be greater slightly further away (1–2 km distance) than further in (< 1 km). It is not valid, however, to make up *post hoc* 'just so' stories to explain apparently anomalous results. In this situation, there must be clear independent evidence that wind direction does indeed influence exposure, and that settlement patterns do indeed interact with wind direction in this way.

So-called 'differential misclassification of exposure' is another story, and can lead to excess observed risks where there are in fact none. This comes about if disease status affects the likelihood that people of equal exposure status will be classified as exposed or not. For example, if one relies on the recall of cases and controls in a case-control study, then it is possible that people with disease (cases) try harder to recall exposure events than people without disease (controls). Cases might for example report more recreational activities near the waste site, or consumption of home grown vegetables, or extreme odour episodes. Where exposure is measured simply as distance of residence from the site, differential exposure misclassification is unlikely.

Whereas it is common knowledge that poor exposure assessment is the Achilles heel of environmental epidemiology, it is less clear how to move forward. Of course, one would attempt to move up the exposure hierarchy mentioned above, and use of biomarkers of exposure is commonly seen as a desirable way forward. Apart from the expense, however, there are other problems. The range of chemicals detected as emanating from landfills is vast[3,15] and expanding.[16] Information about potential toxicity or teratogenicity is often incomplete and it is not clear whether the mixture itself is relevant rather than assessing the

[15] G. Eduljee, Assessment of risks to human health from landfilling of household wastes, in *Risk Assessment and Risk Management*, ed. R.E. Hester and R.M. Harrison, *Issues in Environmental Science and Technology*, No. 9, Royal Society of Chemistry, Cambridge, UK, 1998.
[16] J. Feldmann and A.V. Hirner, Occurrence of volatile metal and metalloid species in landfill and sewage gases, *Int. J. Environ. Anal. Chem.*, 1995, **60**, 339–359.

components independently.[3,17] The relative probability of different pathways of exposure, through groundwater, surface water, air, airborne dust is generally little known, or not in a form useful for epidemiological study of large populations. Some studies of a limited number of sites have measured the levels of chemicals in blood and urine of populations near waste sites and incinerators or level of chemicals in soil, plants, house dust or indoor air[10,18-23] but these measurements have not been directly related to health status other than to guideline levels.

Notwithstanding the problems mentioned above, it might be helpful for future epidemiological studies of large populations if exposure or biomarker studies on population samples near a wide variety of landfills could help to determine the main determinants of exposure of nearby residents (*e.g.* in terms of history and characteristics of the site and its management, climatic conditions, distance from site, location of residence in relation to wind, groundwater flow, and transport of waste, recreational activities, water source, and any other potentially relevant factors). Ideally, the most important determinants could then be measured in an epidemiological study relating exposure to health outcomes, since it would be too expensive to make a total exposure assessment for all residents. Exposure studies could also help to determine whether exposure to a range of chemicals may be sufficiently correlated that it is only necessary to measure selected 'indicator chemicals' in epidemiological studies. Studies of exposure tend to be very detailed in relation to specific sites, often on or near the site boundary, but difficult to extrapolate to larger populations, and are focused on absolute rather than relative exposure measures. For epidemiological surveillance, it is even more necessary to find key surrogates of exposure which can be incorporated into routine data collection.

A final aspect of exposure assessment that needs a mention is the difference between when the exposure occurred and when the exposure was 'generated'. Landfills for example may cause contamination for many years to come. Practices regarding containment may improve, but contamination may remain due to previous practices at the same site. The results of a study therefore do not

---

[17] B. Johnson and C. DeRosa, Chemical mixtures released from hazardous waste sites: implications for health risk assessment, *Toxicology*, 1995, **105**, 145–156.

[18] T. Pless Mulloli, B. Schilling, O. Papke, N. Griffiths and R. Edwards, Transfer of PCDD/F and heavy metals from incinerator ash on footpaths in allotments into soil and eggs, *Organohalogen Compounds*, 2001, **51**, 48–52.

[19] G. B. Hamar, M. A. McGeehin, B. L. Phifer and D. I. Ashley, Volatile organic compound testing of a population living near a hazardous waste site, *J. Expo. Anal. Env. Epidemiol.*, 1996, **6**, 247–255.

[20] J. S. Reif, T. A. Tsongas, W. K. Anger, J. Mitchell, L. Metzger, T. J. Keefe, J. D. Tessari and R. Amler, Two stage evaluation of exposure to mercury and biomarkers of neurotoxicity at a hazardous waste site, *J. Toxicol. Environ. Health*, 1993, **40**, 413–422.

[21] P. A. Stehr-Green, W. W. Burse and E. Welty, Human exposure to polychlorinated biphenyls at toxic waste sites: investigations in the United States, *Arch. Environ. Health*, 1988, **43**, 420–424.

[22] G. F. S. Hiatt, J. Cogliano, R. A. Becker, D. M. Siegel and A. Den, Vinyl Chloride Action Levels: indoor air exposures at a Superfund site, in *Hazardous Waste and Public Health*, International Congress on the Health Effects of Hazardous Waste, ed. J.S. Andrews, H. Frumkin, B.L. Johnson, M. A. Mehlman, C. Xintaras, J. A. Bucsela, Atlanta, Georgia, 1994.

[23] Institute for Environment and Health, *Nantygwyddon landfill site, Gelli, Rhondda – exposure and toxicological review*, University of Leicester, October 2001.

necessarily relate to the risks of waste management practice today, but they do relate to the risks to the population today related to waste management in the past. To study waste management practice today requires choosing sites not previously used and leaving a lag period to allow contamination to build up and to allow a sufficient study population to be 'exposed' to the new practices.

5   Confounding

To be a confounder, a factor must be a risk factor for the disease and it must be associated with the exposure of interest in the study population. Confounding can lead to either overestimation or underestimation of risk related to the exposure, depending on how the confounding factor is related to both disease and exposure. In environmental epidemiology, one of the main types of confounding of concern is socio-economic confounding. Socioeconomic status is associated with the risk of a wide range of diseases and health status itself. Socioeconomic status may also be associated to the exposure in the study population. For example, communities near waste management sites may be less affluent because the undesirable aspects of the presence of the site depresses property prices in the area encouraging poorer people to move in, or because more deprived communities are less able to mount sufficient opposition to a site being established. The socio-economic profile varies with distance from a waste disposal site depending on the history and characteristics of the site and local population, and also depending on other industries and economic influences. Whatever the overall relationship with socio-economic deprivation, when considering any one site it may well be that the more distant population is situated in a densely populated industrial deprived area, and thus the population near the landfill is relatively less deprived. A British study of landfills[14] found an overall tendency in England for the 80% of the population living near open and closed landfills to be more deprived than the remaining 20% of the population. The EUROHAZCON study found that the pattern of variation of socio-economic profile with distance varied from site to site, and there was no overall strong tendency for those near sites to be more deprived than those further away.[6]

If confounding is strong, adjustment for socio-economic variables may not be sufficient, leaving 'residual confounding'. The potential for residual confounding must be assessed for each study separately. In the EUROHAZCON study, residual socio-economic confounding was unlikely to explain the excess risks found. Firstly, statistical adjustment for measured confounding had little effect on the risk estimates, suggesting either that there was little confounding, or that the measurements of socio-economic status were virtually uncorrelated to 'true' socioeconomic status. Secondly, as mentioned above, no overall strong pattern for deprivation to be greater near to the study landfills emerged. Thirdly, when results for non-chromosomal and chromosomal anomalies were compared, the estimates of excess risk were very similar.[24] Chromosomal anomalies such as

---

[24]   M. Vrijheid, H. Dolk, B. Armstrong, L. Abramsky, F. Bianchi, I. Fazarinc, E. Garne, R. Ide, V. Nelen, E. Robert, J.E.S. Scott, D. Stone and R. Tenconi, Chromosomal congenital anomalies and residence near hazardous waste landfill sites, *Lancet*, 2002, **359**, 320–322.

Down Syndrome rather uniquely have a negative association between risk and socio-economic status due to the intermediary maternal age, more affluent communities tending to have a higher average age of mother at delivery than more deprived communities and thus a higher prevalence rate of Down Syndrome. Residual socio-economic confounding would lead to an apparently protective effect of living near landfills for chromosomal anomalies.

Those interested in environmental justice are quite reasonably seeking not to see socio-economic status as a confounder, but as the principal variable of interest in relation to the health impact of waste management. If one is seeking to determine the extent to which landfills impact on health independently of socio-economic status, then it is appropriate to control for socio-economic confounding. If one is seeking to demonstrate the extent to which different socio-ecomic groups suffer a greater or lesser burden from exposure to waste-related hazards, then it is appropriate to present the results in a different way. The two arguments are complementary, not conflicting. There is also a third possibility *i.e.* that the impact of landfills on health varies according to socio-economic status. For example, more deprived communities may have greater exposure due to lifestyle characteristics such as where they spend recreational time, time spent outdoors, food or water source, or in some countries whether they comb the landfills for items to sell or recycle. More deprived people may also be vulnerable in other ways, for example their nutritional or immunity status may be such that they are more vulnerable to chemical exposures or exposure to infection. Thus, it may be appropriate to look for interaction between socio-economic status and measured exposure to landfill in their impact on health.

A second source of confounding is occupational. Do those living close to a waste disposal or treatment site work in more hazardous occupations, either within waste management or another sector? A third source is residence near other industrial sources – do waste management facilities tend to be sited near other potentially hazardous industries? Either of these sources of confounding are possible but not obvious. Waste management tends not to be a big employer, especially of women (in relation to pregnancy outcomes), and occupational risks would need to be very high to influence community risk substantially. It remains to be conclusively demonstrated that the risk of congenital anomaly is raised near any type of industrial site at all, whether waste management or not. It is therefore not a strong argument to suppose that apparent risks near landfills must surely be due to some other sort of industrial site. Nevertheless, a study by Marshall *et al.* in the United States[25] which looked both at landfills and proximity to other industrial facilities found that the strongest evidence related to residence near solvent emitting industry rather than landfill. In that study, the relationship between the two congenital anomalies studied (central nervous system and musculoskeletal) with residence near landfills in general was not significantly elevated after controlling for socio-demographic factors, but there was some suggestion though not directly presented in the results that a greater proportion of the population near landfills also lived near solvent emitting industries.

[25] E.G. Marshall, L.J. Gensburg, D.A. Deres, N.S. Geary and M.R. Cayo. Maternal residential exposure to hazardous wastes and risk of central nervous system and musculoskeletal birth defects, *Arch. Environ. Health*, 1997, **52**, 416–425.

Certainly the way forward on this issue would seem to be to fund studies to look across a classified inventory of all potentially polluting sites in the study population. There has been reluctance on the part of government to sponsor such activity, I believe mainly because of a fear that multiple testing of associations between a wide range of industrial pollution sources and wide range of health outcomes will generate many spuriously statistically significant results (*i.e.* chance associations) which will be difficult to interpret and will create unjustified public concern. In a properly designed system of envirovigilance, or environmental health surveillance, we will need to find a way through this impasse. Another relevant issue however is whether to invest resources in large scale (in population terms) but broad brush studies largely based on proximity to polluting sites, or whether to invest in more detailed studies of the potential for exposure and biological effect, or both.

When the results of a study such as EUROHAZCON are received, attention turns to an even wider range of possible confounders. Might intake of teratogenic drugs, maternal infections during pregnancy, alcohol, smoking, nutrition and other risk factors differ in populations near sites? It is difficult to see why these factors should be particularly prevalent near waste disposal sites, other than through their relationship with socio-economic status, which has been controlled for. It is possible that 'risk taking' behaviours might group together and that they include choice of living environment and attitude to proximity to industrial sites, but there is little research on this. In the case of congenital anomalies, known risk factors explain only a small proportion of cases. It may be more important to invest in detailed exposure assessment than detailed confounder assessment. In particular, the finer and more accurate the exposure assessment, the less likely that gradations of exposure will correlate strongly with confounding factors.

Studies of single waste disposal sites are open to more potential confounding than multisite studies when distance of residence is used as the main exposure measure. In multisite studies, there would need to be a general relationship between the confounder and the exposure, *e.g.* people near sites take more teratogenic drugs during pregnancy than people further from sites, or landfills tend to be associated with sources of industrial pollution with high teratogenic potential in their vicinity. For single site studies, however, there does not have to be a general relationship between the exposure and confounder: it may just be a chance association. For example, local medication or diet may be unusual, or it may be the case that the landfill is close to a massive industrial site in this particular area, leading either to community or occupational exposures. Confounding is therefore more likely to arise in single sites studies and, in this respect, multisite studies are more convincing evidence of an impact of waste disposal sites on health.

A potentially powerful device to control for confounding is to try to integrate a 'natural experiment', *e.g.* comparing the populations before and after the beginning of site operation. This has been done in a number of studies of waste management),[14,26–28] sometimes leading to stronger and sometimes weaker

[26] H. M. P. Fielder, M. Poon-King, S. R. Palmer, N. Moss and G. Coleman, Assessment of impact on health of residents living near the Nant-y-Gwyddon landfill site: retrospective analysis, *Br. Med. J.*, 2000, **320**, 19–23.

evidence of a causal association. It is only possible if health data go back to before site operation. Interestingly, following the study of the Welsh Nant-y-Gwyddon site where it was demonstrated that excess congenital anomaly risks were present in the area both before and after the start of landfill site operation, the local community protested that this was because before site operation, forestry practices in the area had led to high pesticide exposure (personal communication). While one is tempted to feel that this is not a 'parsimonious' explanation, it nevertheless points to the relative simplicity of epidemiological assessments in relation to the complexity of the situation pertaining to a single landfill site.

6 Statistical Considerations

Generally, the rarer the 'disease', the rarer the exposure, and the smaller the risk among the exposed relative to that among the unexposed, the greater will be the population sample size needed to have a study of adequate statistical power. Choice of appropriate epidemiological study designs are on the one hand about minimising bias, and on the other hand about maximising statistical power for a given number of study subjects. For example, for a rare outcome such as a specific congenital anomaly, one might choose a case-control study to maximise statistical power for a given number of study subjects, and for a rare exposure one might choose a cohort study for the same reason. The most difficult situation is where both disease and exposure are rare, and the relative risk is low. The larger the study in terms of number of study subjects needed, the more expensive it is liable to be, and the more likely that one will need to save money by foregoing detail in disease or exposure classification, and that one will lose control over the consistency of disease or exposure measurement across the entire study population.

A device commonly used to increase statistical power is to group different diseases or exposures together (thereby making them less rare). Congenital anomalies as a whole, for example, affect approximately 2% of all births. Specific anomalies may affect one in a thousand births (*e.g.* neural tube defects) or one in ten thousand (*e.g.* gastroschisis). Since little is known about the aetiology of the majority of congenital anomalies, it is not always clear whether and how to group different anomalies together.[29] By grouping, one might miss risks confined to specific types of congenital anomaly. There is an optimal balance between gaining power by grouping and losing power by lowering the overall relative risk, but this cannot be known in advance.

Another device used to increase statistical power is to find highly exposed groups where the relative risks to be detected will be highest. Occupational exposures are usually higher than general community exposures. There have been some studies of waste management facility workers,[3,30–32] but it is generally

[27] M. Berry and F. Bove, Birth weight reduction associated with residence near a hazardous waste landfill, *Environ. Health Perspect.*, 1997, **105**, 856–886.

[28] S. H. Swan, G. Shaw, J. A. Harris, R. R. Neutra, Congenital cardiac anomalies in relation to water contamination, Santa Clara County, California, 1981–83. *Am. J. Epidemiol.*, 1989, **129**, 885–893.

[29] H. Dolk, The role of the assessment of spatial variation and clustering in the environmental surveillance of birth defects, *Eur. J. Epidemiol.*, 1999, **15**, 839–845.

[30] A. Hartmann, H. Fender and G. Speit. Comparative biomonitoring study of workers at a waste disposal site using cytogenetic tests and the comet (single cell gel) assay, *Environ. Mol.*

a small workforce and biomarker-based small studies are of interest in this group. Workers, however, differ in age and sex distribution from the general population, and are also self-selected to be relatively healthy, so care is needed in extrapolating results to the general population. Moreover, the question still remains of how to extrapolate findings to the community with lower, and perhaps qualitatively different, exposures.

Many communities would like to know the impact of their local waste management facility on health. A single facility study is likely to lack statistical power for the rarer outcomes. Multisite studies can enhance statistical power. However, it is not always clear how applicable the general results are to the local site, particularly when a heterogeneous group of sites has been studied. The temptation is to engage in numerous subgroup analyses to try to get back to the characteristics of the local site in isolation. This is rarely a worthwhile enterprise unless foreseen in the design of the study. Subgroup analyses lead on the one hand to the loss of statistical power so that results are non-significant with large confidence intervals around point estimates, and rather uninformative. They also lead to the problem of multiple testing, where some nominally statistically significant results are found due to the play of chance. For example, in every 100 statistical tests done, 5 will be 'statistically significant' at the commonly employed 5% significance level. It is important to choose subgroup analyses carefully on the basis of prior hypotheses about the nature of differences, rather than embark on a 'fishing expedition', and more powerful still to look for evidence of dose–response, *i.e.* a graded relationship between exposure dose and health outcome (see Criteria for Causation below).

When reviewing the literature on single site studies, consideration of publication bias is relevant. This can occur in two stages. A local community may be concerned about an apparently high rate of a certain disease, and seek the nearest reason, perhaps a waste disposal site or a power line. Attaching a *p*-value to this association between exposure and disease (the probability that this would arise by chance) is invalid and known as the 'Texan sharpshooter problem'. The Texan draws his gun and fires randomly at the barn door, then goes to the barn door and draws a target around the densest collection of bullet holes, claiming he is a good shot. In other words, as there are many communities, and a random distribution of disease will produce chance aggregations in some places, and as there are many landfills, some disease aggregation–landfill associations will arise by chance and one cannot draw a target around these for evaluation of statistical significance. The second stage in publication bias is that, even if studies of single landfills did not selectively choose communities with disease aggregations, scientific journals are not open to reporting one by one many negative associations between disease and landfills or other sites. Positive associations have a better chance of publication. Multisite studies avoid these types of bias. They predefine a set of

Mutagenesis, 1998, **32**, 17–24.

31 M. E. Gonsebatt, A. M. Salazar, R. Montero, F. Diaz Barriga, L. Yanez, H. Gomez and P. Ostosky-Wegman, Geonotoxic monitoring of workers at a hazardous waste disposal site in Mexico, *Environ. Health Perspect.*, 1995, **103** (suppl. 1), 111–113.

32 H. Fender and G. Wolf, Cytogenetic investigations in employees from waste disposal sites, *Toxicol. Lett.*, 1998, **96–97**, 149–54.

Figure 2 Bradford Hill's
Criteria for Causality[33]

Strength of association (a high relative risk is less likely to be explained by bias)
Consistency (in different populations under different circumstances)
Specificity (a cause leads to a single effect)
Temporality (cause precedes the effect in time)
Biological gradient (presence of a dose–response effect)
Biological plausibility
Coherence (between different types of evidence)
Experimental evidence (where only the factor of interest is varied)
Analogy

landfills or incinerators irrespective of disease status of the local populations, and likelihood of publication does not depend on whether the results are positive or negative (although choice of journal may).

7   Criteria for Causation

It should be clear from what has been discussed above that no one observational epidemiological study should normally be interpreted as 'sufficient proof' or even strong evidence of a link between a certain waste management practice and an adverse health effect. Association (the association of the risk factor and the disease in the same population subgroups) does not necessarily mean causation (that the risk factor has a part to play in causing the disease). In 1965, Austin Bradford Hill[33] came up with a series of standards by which one can assess the strength of evidence for causation from epidemiological evidence for association (Figure 2), still in use today. While these criteria can be considered for the interpretation of a single study, they are meant for the interpretation of the body of evidence including all available studies. The criterion 'strength of association' is rarely satisfied in environmental epidemiological studies as one expects a relatively small effect in terms of relative risk, at least compared to occupational exposures. In fact, when a 'cluster' of disease is found near a landfill or incinerator, such that the risk of disease seems to be tenfold higher than expected, this unusually scores high on the 'strength' criterion, but it may score low on the 'biological plausibility' criterion. Consistency and coherence are very important, as they refer to consistency of results across studies, and different types of evidence pointing the same way, in particular suggesting that bias and chance are not likely explanations. Specificity is useful again to eliminate the possibility of bias, which would be less likely if the association can be shown to relate only to one exposure–disease relationship and not others. However, it can also be expected that exposures will lead to a range of diseases, and in fact it is extremely difficult to judge this in the absence of more aetiologic information about the diseases being investigated. Dose–response or biological gradient is one of the main criteria sought in epidemiological study design. For example, the EUROHAZCON study sought to develop a hazard potential ranking to compare the relative risk near the more hazardous sites to that near the less hazardous sites. This approach of course depends on the validity of the ranking

[33] A. Bradford Hill, The Environment and Disease: Association or Causation?, *Proc. R. Soc. Med.*, 1965, **58**, 295–300.

method, and the external validity of the hazard ranking method was in this case difficult to establish.[11] Distance itself can be seen as a form of dose–response, if it can be demonstrated that risk varies as a function of graded distance, rather than simply when distance is dichotomised to 'near' and 'far'. Finally the criteria biological plausibility and analogy have been useful to assess causality. For example, rubella and thalidomide were the first environmental causes of congenital anomalies to be established, and opened the way to investigating environmental causes in general. There are many instances in epidemiological history however where biological plausibility followed many years after a clear association between exposure and disease was established.[33]

Bradford Hill's criteria can also be useful in assessing where to invest further research resources. For example, it may be relatively uninformative to repeat on an ever larger scale studies which essentially suffer from the same sorts of bias, rather than investing in more varied design approaches. It may also be a waste of time to analyse a single database more intensively, with all the attendant problems of subgroup analyses above, rather than focusing on consistency across different approaches. It may also be fruitful to look at biological measures to explore biological plausibility.

8   Use of Epidemiological Studies in Risk Management

The Bradford Hill criteria or their intuitive equivalents are not widely employed by the media and public, and often not even by scientists or policy makers. Each individual epidemiological study appears in publication with its caveats (a discussion of possible sources of bias). On the one hand, those who wish to believe that the results imply a causal connection choose to ignore these caveats. On the other hand, those who wish not to believe that the results imply a causal connection choose to use the caveats as invalidation of the study results. Both of these approaches avoid recognition that caveats are in the nature of single observational epidemiological studies, and that the important thing is to build up evidence from multiple studies of differing designs and populations.

In the real world however, it is not possible to 'sit on the fence' with epidemiological caveats. Decisions about new or extended landfills and incinerators need to be made now. Should landfills and incinerators by allowed near populated areas? How near? With what restrictions? Any scientific literature review will conclude that there is considerable scientific uncertainty and this is of course a central problem in using risk assessment for risk management, and led to adoption of the 'Precautionary Principle'.

The public has become more aware that it does not live in an environment of guaranteed safety, where all risks have been measured and assessed and agreed to be tolerable. On the other hand, some argue that the public have become less tolerant of risk. We need informed debate about which risks we are prepared to live with and who should bear them. Further investment in research and surveillance on the health impact of waste management should inform this debate. Deciding where to invest our research resources, however, requires a broader look at how different strategies of balancing production and waste and different strategies of waste disposal may impact on health.

Subject Index

Abdominal wall defects, 115, 121
Acetogenesis, 143
Acetogenic leachates, 149
Acid gas, 155
Actinomycetes, 81
Activated carbon, 158
Acute and chronic CR functions, 182
Adjuvants, 86
Adverse birth outcomes, 103, 116
Aerated static piles, 74
Air pollutants, classical, 174
Air pollution control, 163
Allergenic fungi, 80
Allergic asthma, 84
Allergic rhinitis, 84
Alternaria spp., 84
Ambient air quality standards, 182
Ambient bioaerosol levels, 91
Arsenic, 185
Ash residues, 163
Aspergillosis, 80, 99
Aspergillus spp., 78
Aspergillus fumigatus, 79
Asthma, allergic, 84
Atmospheric dilution, 131
Atmospheric modelling, 131

Bacillus spp., 78
Bacteria, 80
  coliform, 79
  psychrotrophic, 89
Bacteria and fungi (bioaerosols), 55
Bioaerosols, 55, 63, 79
  concentrations, 65
Biodegradation, 142
Biological markers, 199
Biological plausibility, 209
Biomarkers of exposure, 202
Bioreactive waste, 88, 142, 143
Birth defects, 103, 115, 116, 121, 196, 207
  chromosomal, 118
  non-chromosomal, 117

Birth weight, low, 103, 115, 116, 121
BOD, 148
Bottom ash, 163, 165, 172
Bradford Hill, Austin, 209
Bronchitis, 82, 85

Cadmium, 67, 71, 155, 158, 185
Calorific value, 153
Cancer,
  incidence rates, 109, 121
  risks, 131, 136
Carbon, activated, 158
Carbon dioxide, in landfill gas, 131, 144,
    145, 146
Carbon monoxide, from waste
    incineration, 184
Carcinogens, 80
Cardiac septa, 118
Case-control studies, 104
Causal connection, 210
Cement kilns, 177
Central sorting, 15
Chromium, 185
Chromosomal birth defects, 118
Chronic bronchitis, 85
Chronic health effects, 66, 183
Chronic obstructive pulmonary disease, 85
Cladosporium spp., 84
COD, 148
Cohort studies, 104
Coliform bacteria, 79
Colony forming units (cfu), 79
Combustion, 147, 155
Composting, 10, 17, 165, 166, 167
Compost workers, 84
Congenital anomalies, *see* Birth defects
Contingent valuation, 185
Criteria for causation, 209

de novo synthesis, 160
Development effects, 134
Diarrhoea, 99

Diesel exhaust particles, 71
Dilution factors, 131, 138, 175
Dioxins, 23, 147, 148, 150, 155, 160
Direct contact exposure, 129
Disease registers, 198
Dispersion models, 179
Dose–response function, 172, 208, 209, 100
Down's syndrome, 117, 119
Dust, 63, 87, 129

EcoSense software, 179
Electromagnetic fields, 67, 71
Embryotoxic effects, 126
EMEP model, 179
Emission limits, 156
Endotoxins, 55, 68, 81
Energy recovery, 177
Environmental assessment levels, 135
Environmental concentrations, 64
Environmental epidemiology, 196
Environmental fate, 172
Environmental health surveillance, 206
Environmental justice, 205
Envirovigilance, 197, 206
Epidemiological studies, 103, 104, 109,
    196, 203, 206
ESR, 71
EU Directives,
    Waste Incineration, 155, 156, 162,
    171
    Waste Landfill, 53, 72, 152
EUROHAZCON, 109, 117
Exposure, 91, 201
    assessment, 199
    determinants of, 203
    pathways, 129
    risk assessment studies, 103
ExternE project, 172
Extrinsic allergic alveolitis, 81, 85

Fabric filters, 161
Farmers' lung disease, 81
Fetotoxicity, 126
Fluidised bed combustion, 25
Flyash, 158, 163, 164, 172
Food chain, 200
Furans, 147, 148, 150, 155, 160

Gases,
    emissions, 103, 129, 131
    greenhouse, 146, 174

landfill, 30, 126, 129, 131, 135, 144, 145,
    146
migration, 109, 129
Gasification, 167, 168
    and pyrolysis, 26
Gas migration, 109, 129
Gastric infections, 99
Gastroschisis, 116
Geographical studies, 104
Gestational periods, 127
Global warming, 196
Glucans, 68
    $(1 \rightarrow 3)$-β-D-glucan, 55, 83
Granulomatous pneumonitis, 85
Greenhouse gases, 146, 174
Green waste, 87
Groundwater, 200

Harwell Trajectory model, 179
Hazardous waste landfill, 196
Hazard potential ranking, 200
Health criteria, 129, 135
Health effects, 66, 67, 69, 183
Health impact assessment, 195
Health of workers, 54, 66
Health status measurement, 197
Heat recovery, 63, 178
Heavy metals, 152, 153, 157, 158, 164, 167
Home composting schemes, 10
Household wastes, 54, 126
Hydrogen chloride, 159
Hydrogen fluoride, 159
Hydrogen sulfide, 144
Hypospadias/epispadias, 115, 121, 122

IgG antibodies, 85
Immunoglobulin E (IgE), 84
Immunotoxicity, 199
Incineration, 72, 153
    EU Directive, 155, 156, 162, 171
Industrial wastes, 126
Inequalities between populations, 195
Infection, 99
Ingestion, 79
Inhalable dust, 87
Inhalation exposure, 79, 103, 129
Integrated waste management, 34

Klebsiella spp., 89

Landfill, 28, 72, 103, 120, 141, 171

bans, 32
design and operation, 29
EU Directive, 53, 72
gas, 30, 126, 129, 131, 135, 144, 145, 146
gas recovery, 24
hazardous waste, 196
Leachate, 142, 147, 148, 149, 150, 151, 166
plume, 152
Lead, 167
Life cycle analysis, 169
Life cycle assessments, 189
Lipopolysaccharides, 81
Lixiviation, 172
Long life products, 12

Markers of biological effect, 198
Materials recovery/recycling facilities, 53
workers, 54
Maturation (curing), 76
Mercury, 67, 71, 155, 158, 167
Metal concentrations, 65
Methane, 131, 144, 145, 146
Methanogenesis, 144, 149
Micropollutants, 175
Mixed domestic waste, 72
Mixed waste materials recovery facilities, 15
Mixed waste sorting plants, 58, 60
Monetary valuation, 185
Monocytes, 70
Multi-site epidemiological studies, 104, 109, 206
Municipal biowaste sites, 88
Municipal solid waste, 2, 4, 120, 126
Mushrooms, 90
Mushroom workers lung disease, 81
Mycotoxins, 80, 82

Neural tube defects, 115, 118, 121, 122
Nickel, 185
Nitrates, 185
Nitrogen oxides, 147, 159, 160, 184
Noise, 63
Non-chromosomal birth defects, 117
Non-special sites, 121
NO_x, 147, 159, 160, 184
Nutritional factors, 115

Occupational exposure, 207
limits, 56
Odds ratio, 118

Odour, 126, 127
Organic compounds, volatile, 67, 71, 83, 126
Organic dust toxic syndrome, 56, 60, 82, 85
Organic substrates, 73
Ozone, 184, 98

Paper, recycled, 63, 169
Particulate matter, 66, 156, 183
diesel exhaust, 71
PM_{10}, 157
total suspended, 64
Pathogens, 89
PCBs, 150
and pesticides, 64
PCDD, 160, 161, 164
PCDD/F, 161
PCDF, 160, 161, 164
Penicillium spp., 78, 84
Pollen, 93, 98
Pollutants,
primary, 174
secondary, 174, 175
Prematurity, 126
Prenatal care, 126
Primary pollutants, 174
Product design, 12
Product durability, 12
Psychrotrophic bacteria, 89
Publication bias, 208
Public attitudes, 43
Pyrolysis, 167, 168, 177

Quality of life, 198

Ragweed pollen, 98
Reclamation, 31
Recycling, 33, 63, 168, 169
Reduction of waste volume, 172
Refuse-derived fuel (RDF), 25, 58
Relative risk, 121, 123
Rhinitis, allergic, 84
Rhizopus spp., 84
Risk assessment, 94, 129
studies, 103
Risk management, 210

Saccharopolyspora (*Faenia*) *rectivirgula*, 85
SAHSU Study, 109, 121
Salmonella typhimurium, 89

Screening, 75
Scrubbing systems, 159, 161
Secondary air pollutants, 174, 175
Sensitisation, 96
Sewage, 90
Shredding, 74
Single-site epidemiological studies, 104, 109
Slag, 172
Socio-economic confounding, 204
Socio-economic deprivation, 121
Socio-economic status, 115, 205
Soil,
    gas migration, 109
    micro-organisms, 92
Source-separated MRFs, 17
Special (hazardous) wastes, 121
Spores, 93
Stabilisation in concrete, 172
Statistical artefacts, 123
    confounding, 122
Statistical power, 104, 119, 207
Still births, 121
Stress, 197
Sulfates, 185
Sulfur oxides, 159
Superfund program, 129
Symptoms, 66
Synergistic effects, 131
Synthesis, *de novo*, 160

'Texan sharpshooter problem', 208
Thermal treatment, 22
Thermoactinomycetes spp., 85
Thermophilic actinomycetes, 78
Thermoselect system, 168
Thermus spp., 78
Thresholds, 173

Total suspended particulates, 64
Toxic pneumonitis, 85
Trace gas constituents, 131, 135
Trends in waste management, 39, 43

Uniform world model, 187
Unit risks of cancer, 136
Unsorted waste, 61
Urbanisation, 122

Value of statistical life, 185
Vermicomposting, 75
Verocytotoxic *E. coli* O157, 80
Volatile organic compounds, 67, 71, 83, 126

Waste,
    handling, 63, 65
    household, 54, 72, 126
    industrial, 126
    minimisation, 9, 11, 12
    mixed, 15, 58, 60, 72
    municipal, 2, 88, 120, 126
    sorting facilities, 66
    special (hazardous), 121
    strategy, 53
    unsorted, 61
Waste management, 58
    hierarchy, 3
    integrated, 34
    trends, 39, 43
Waste management facility workers, 208
Waste recycling worker syndrome, 66
Wastewater, 161, 162
WHO guideline values, 136, 182
Windrow systems, 74

Years of life lost, 186